Using Your Chemistry

Comprehension Questions for Advanced Level

P R Scott M.A., D.Phil.
Charterhouse, Godalming

Stanley Thornes (Publishers) Ltd

First published in 1990 by:
Stanley Thornes (Publishers) Ltd
Old Station Drive
Leckhampton
CHELTENHAM GL53 0DN
England

British Library Cataloguing in Publication Data
Scott, Peter
   Using your chemistry: comprehension questions for
   advanced level.
   1. Chemistry
   I. Title
   5400

   ISBN 0–7487–0437–X

The front cover photograph shows a model of a zeolite; three cages are occupied by Na^+ ions and the fourth contains an alkane molecule. The photograph was kindly supplied by BP Research Ltd.

Typeset in Palatino and Helvetica by Tech-Set, Gateshead, Tyne & Wear.
Printed and bound in Great Britain at The Bath Press, Avon.

Contents

Preface

To the student

You may find the questions in this book rather unfamiliar, and indeed you may think they are quite difficult at first. They do not try to test you on situations that you have already seen and thought about; instead they give you information on topics that you have not studied before. The object is to allow you to read new material with understanding, applying what you have done before; that is, after all, the way you will have to use your chemical knowledge in the future.

You will need to read the passages carefully; remember, they contain the clues necessary for your answers. Think also about your answers to other questions; sometimes a series of questions is designed to lead you through an argument, with each step building on a previous answer. Lastly, have an eye on the mark scheme when you write your answers; one mark will not require an essay, but if there are three marks, you should think whether you have made three separate points. Although you may not feel like it at the moment, I hope you enjoy the exercises.

To the teacher

In recent years there has been a steady move in examining chemistry at A-level away from testing a large body of factual material and towards asking candidates to apply their knowledge. Comprehension exercises are being steadily introduced, and this book is written in response to the need for candidates to have experience of these questions.

The book is divided into three parts: comprehension exercises, data analysis questions, and a series of tests on experimental situations. Many of the tests will be used towards the end of an A-level course, but there are a number which should be suitable for candidates in their first year.

The questions have been pre-tested, which has helped to show up some ambiguities in what was intended, but I have no doubt that other perfectly reasonable misinterpretations are still to come. A book containing the answers that I had in mind is available separately.

P.R.S.

1 Reprocessing Nuclear Fuel

Most nuclear reactors operate by the fission of ^{235}U nuclei by neutrons. When a ^{235}U nucleus is struck by a relatively slow-moving neutron it splits, emitting energy and forming smaller nuclei and further neutrons. These neutrons can then split more ^{235}U nuclei, and by controlling the rate of absorption of neutrons, the rate of fission can be controlled. Naturally occurring uranium consists of two isotopes, ^{235}U and ^{238}U, with the former making up only about 0.7%; the ^{238}U isotope is not 'fissile', but it is 'fertile', as it can absorb neutrons, eventually producing fissile Pu atoms.

As a reactor operates the ^{235}U atoms are slowly used up; when the fissile content becomes too low, the fuel rod is replaced. The used fuel rod typically contains about 99.2% U, 0.5% fission products and 0.3% Pu. Although the fission products are not useful, the U and Pu are valuable, and are recovered.

The fuel rods are left to stand in cold water ponds for six months, and are then dissolved in 7M nitric acid. The uranium and plutonium are converted into $UO_2(NO_3)_2$, uranyl nitrate, and $PuO_2(NO_3)_2$. These two compounds are soluble in a mixture of tri-butyl phosphate (TBF) and kerosene, and are removed from the aqueous layer by solvent extraction. The fission products form nitrates which remain in the aqueous phase; the water is then removed from this phase, and the remaining material is stored in high-integrity steel tanks.

The kerosene phase is now treated with a solution of iron(II) sulphamate in water, which converts the plutonium to Pu^{3+} ions. These are insoluble in the organic phase, and the Pu returns to the aqueous phase. It is eventually converted to PuO_2. The uranyl nitrate is now concentrated by evaporation, and then thermally denitrated to form UO_3. The plutonium and uranium can both be used in fast breeder reactors to give more energy.

---

1 Give one difference (other than the composition of the nucleus) between ^{235}U and ^{238}U. [1]

2 Explain in your own words the terms fissile and fertile (para 1). [2]

3 Why do you think the used nuclear fuel is stored under water for six months before being reprocessed? [2]

4 a) Given that uranyl nitrate contains NO_3^- ions, calculate the oxidation number of uranium in it. [1]
   b) What other ion does uranyl nitrate contain? [1]
   c) Explain why you would not expect uranyl nitrate to dissolve in kerosene alone. [1]

5 What role does the iron(II) sulphamate play? [1]

6 Why is the remaining material stored in high-integrity steel tanks? What might be done with it after that? [3]

7 What can you say about the solubility of (a) water in TBP–kerosene, (b) uranyl nitrate in TBP–kerosene and (c) plutonium(III) nitrate in water? [3]

8 What can you say about the relative ease of conversion of uranyl nitrate into uranium(III) nitrate, and the corresponding reaction with plutonium? [1]

9 What do you understand by the term 'is thermally denitrated to UO_3'? Suggest an equation for the reaction. [2]

10 In what way would the uranium recovered from the processing of nuclear fuel differ from naturally occurring uranium? [2]

2 Swimming Pools

The main health hazard in swimming pools is the growth of bacteria, which are usually introduced to the water by swimmers. A number of different techniques have been used to disinfect water, including the use of ozone, metal ions and ultraviolet light, but the most widespread is the use of chlorine.

When chlorine is added to water, it rapidly reacts forming two acids:

$$Cl_2 + H_2O \rightleftharpoons HCl + HClO$$

The hydrochloric acid is fully ionised, being a strong acid, but the chloric(I) acid is a weak acid, so an equilibrium is set up:

$$HClO \rightleftharpoons H^+ + ClO^-$$

The position of this equilibrium is important, as it controls the relative proportions of HClO, a good bactericide, and ClO^-, whose effectiveness is much lower. HCl is not a bactericide. Chlorine may be introduced directly as a gas in some large pools, though smaller pools normally use a solution of sodium chlorate(I). The total chlorine concentration is about $2\,mg\,dm^{-3}$.

It is important to control the pH of the pool water carefully; if the water is too alkaline then scale can be produced, blocking filters, while if the solution is too acidic plaster and metal parts are corroded. The pH should be kept in the range 7.4–7.6; it is usually controlled by the addition of hydrochloric acid or sodium hydrogensulphate ($NaHSO_4$) if the pH is too high, or sodium carbonate if it is too low.

---

1 What are the oxidation numbers of chlorine in Cl_2, HClO and HCl? What sort of reaction is that between chlorine and water? [3]

2 Calculate the mass of chlorine that should be added to a pool 10 m × 5 m × 2 m to produce a concentration of $2\,mg\,dm^{-3}$. [2]

3 Explain what is meant by a 'weak acid', and the 'dissociation constant of a weak acid'. [2]

4 The dissociation constant for HClO is $4 \times 10^{-8}\,mol\,dm^{-3}$. Calculate the ratio of ClO^-/HClO present if the pH is (a) 8.0 and (b) 7.4. Hence explain why the pH should be kept below about 7.6. [3]

5 Explain why the addition of sodium hydrogensulphate decreases the pH of pool water, and why the addition of sodium carbonate increases it. [2]

6 Calculate the hydroxide ion concentration in water of pH 8.0, and in water of pH 7.5 ($K_w = 10^{-14}\,mol^2 dm^{-6}$).

Hence calculate the change in the number of moles of hydroxide ions when a pool $10\,m \times 5\,m \times 2\,m$ has its pH changed from 8.0 to 7.5.

Hence calculate the mass of $NaHSO_4$ needed to bring about the change, assuming that hydrogensulphate ions reacting with hydroxide ions is the only reaction that occurs. (A_r: H 1, O 16, Na 23, S 32)   [5]

7 Your answer in question 6 is in fact a slight underestimate, as there are other chemical changes that occur. Suggest what one might be.   [1]

8 Most swimming pool water tends to become slowly more alkaline, especially if the pool is open to the air, requiring acid to be added. Suggest how this might occur.   [2]

3 High Explosives

A high explosive is a compound which can explode with great violence when detonated by another initiating explosion; its decomposition is exothermic, and is accompanied by the formation of a large volume of gas. This gas causes a sudden local increase in pressure, forming a shock wave. The initiating explosion is caused by an explosive of much greater sensitivity, such as lead azide or mercury fulminate.

Most high explosives are organic nitrates or nitro compounds. The carbon and hydrogen react with the oxygen in the nitrate or nitro groups, forming carbon dioxide and water, along with nitrogen gas. Many explosives do not contain enough oxygen to burn all the carbon and hydrogen; the oxygen balance of an explosive is the number of grams of oxygen lacking or in excess of that needed for the complete combustion of 100 g of explosive. Explosives are often mixed with oxygen-rich compounds, such as chlorates or nitrates, to make up the oxygen balance.

Aromatic nitro compounds are important explosives. When a benzene ring has two or more nitro groups attached to it, then it can be made to explode. Trinitrobenzene is produced by extended treatment of benzene with concentrated nitric and sulphuric acids.

Each stage is more difficult than the previous stage. Trinitrobenzene is a yellow crystalline substance which melts at 123 °C; it is a substance of great explosive strength.

Trinitrotoluene (TNT) is prepared in a similar way, although it is easier to form.

It has less explosive power than trinitrobenzene, but has been used extensively since World War I. It is a solid, but melts at only 81 °C, which makes it much safer to handle when filling cartridges and shells. It contains only a small proportion of the oxygen needed for complete combustion, and so tends to burn with a smoky flame; it is commonly mixed with ammonium nitrate to improve the oxygen balance.

COMPREHENSION

Another explosive compound is picric acid, trinitrophenol. It is formed in a similar way to TNT:

It was largely replaced by TNT at the beginning of the 20th century, but was used again in World War I when supplies of methylbenzene (toluene) were difficult to obtain. Although picric acid is quite stable, it forms metal salts readily, and these are highly sensitive. Picric acid cannot therefore be allowed to come in contact with many metals, nor with lead-containing paints.

1 What is the molecular (not structural) formula of trinitrobenzene? [1]

2 Why is conc. sulphuric acid used in the nitration of benzene rings (para. 3)? [2]

3 Why are methylbenzene and phenol more readily nitrated than benzene itself (para. 4)? [2]

4 Why is each step in the nitration of benzene more difficult than the previous step (para. 3)? [1]

5 Explain in your own words why ammonium nitrate is often added to trinitro-toluene (para. 4). [3]

6 How would you melt trinitrotoluene, in order to minimise the danger of explosion? [1]

7 Give the structural formula of the sodium salt of picric acid. [2]

8 $C_2H_4N_2O_6$, is another explosive. Write an equation for its decomposition, and hence calculate its oxygen balance. [2]

9 Write an equation for the decomposition of trinitrotoluene (ignore the presence of any oxygen in the air, or CO in the products). [2]

10 A mixture is made of trinitrotoluene and ammonium nitrate, such that the oxygen balance is zero, i.e. the only products are carbon dioxide, water and nitrogen. By considering how many oxygen atoms each TNT molecule needs, and how many oxygen atoms each NH_4NO_3 can give, write a balanced equation for the reaction. Hence calculate the percentage of ammonium nitrate by mass in the mixture (A_r: H 1, C 12, N 14, O 16). [4]

4 Synthesis of Natural Gas

There is much interest in processes in which coal is converted into methane gas. Coal consists mainly of carbon, with a typical C:H weight ratio of 16:1. It does not react directly with hydrogen gas, but it will react with steam, forming carbon monoxide and hydrogen, known as 'synthesis gas'.

$$C + H_2O \rightleftharpoons CO + H_2$$

The reaction is endothermic ($\Delta H = +136\,kJ\,mol^{-1}$), but the necessary heat can be supplied by mixing some oxygen with the steam, forming more carbon monoxide; this reaction is strongly exothermic ($\Delta H = -221\,kJ\,mol^{-1}$).

$$2C + O_2 \longrightarrow 2CO$$

The gas produced is too rich in carbon monoxide for conversion to methane, so more steam is added, forming hydrogen in a 'shift reaction' ($\Delta H = -42\,kJ\,mol^{-1}$).

$$CO + H_2O \rightleftharpoons CO_2 + H_2$$

The final stage involves the reaction of carbon monoxide and hydrogen to form methane and water; the reaction is quite exothermic, but needs a nickel catalyst.

$$CO + 3H_2 \rightleftharpoons CH_4 + H_2O$$

The catalyst is formed by heating an intimate mixture of nickel(II) and aluminium carbonates, forming a mixture of the metal oxides; hydrogen is then passed over the oxides, giving finely divided Ni on an aluminium oxide support. The catalyst has a limited lifetime, as its effectiveness is reduced by sulphur compounds, and by whiskers of carbon from the decomposition of carbon monoxide. A typical composition for the mixture obtained after passing gases over the catalyst is:

CO 0.5%	CO_2 51.2%	CH_4 33.1%
H_2 6.2%	N_2 4.0%	others 5.0%

1 What advantages does methane have over coal as a fuel? [2]

2 Why is there interest in the conversion of coal to methane? [1]

3 What is the C:H weight ratio in methane? (A_r: C 12, H 1) [1]

4 What other element present in coal in small amounts might need to be removed before methane could be synthesised in the way suggested? [1]

5 What conditions of temperature and pressure would give the highest yield of carbon monoxide and hydrogen from coal and steam? Explain your reasoning. [4]

6 After the initial stage, the temperature of the mixture is reduced for the shift reaction; suggest why this might be. [2]

7 a) If you were trying to predict the enthalpy change for the shift reaction, using standard tables of heats of formation *or* heats of combustion, what data would you need to look up? (Select one set of tables or the other.) [3]
    b) Why might your answer not be exactly correct for these circumstances? [1]

8 Give balanced equations for the formation of the Ni catalyst. [2]

9 Why should the nickel be finely divided? [1]

10 What is the major disadvantage of this scheme for converting one fuel into another? [2]

5 Neutron Activation Analysis

Neutron activation analysis is a versatile technique for both qualitative and quantitative analysis of chemical elements. The sample under investigation is bombarded with neutrons, which convert stable nuclei into unstable isotopes; these nuclei then undergo radioactive decay, and the products of the decay are monitored. Each unstable nucleus decays with a characteristic half-life, and giving radiation of characteristic energy, so identification of the radioactive nucleus, and hence the original nucleus, is not difficult.

The source of neutrons may be a nuclear reactor, a particle accelerator, or more conveniently an isotope source. Over seventy different elements have at least one isotope that can absorb neutrons, forming radioisotopes; metals in minerals have commonly been investigated using the technique, but it has also been applied to blood plasma and archaeological samples. Gamma rays are the most widely used form of radiation with scintillation counters being used as detectors. A simple example would be the activation of the common isotope of sodium 23:

$$^{23}Na + n \longrightarrow \ ^{24}Na$$

followed by the decay:

$$^{24}Na \longrightarrow \ ^{24}Mg + e^- + \gamma$$

The decay has a half-life of 15 hours, and the γ-rays have an energy of 1.3 MeV.

Neutron activation analysis can be used to analyse the percentage of manganese in an alloy consisting mostly of aluminium, with a few per cent of copper, silicon, magnesium and manganese. The properties of the relevant nuclei are given below:

Isotope	Natural abundance/%	Neutron cross section/b	Half-life	γ-ray energy/MeV
^{24}Mg	79		*	
^{25}Mg	10		*	
^{26}Mg	11	0.03	9.5 min	1.01, 0.84
^{27}Al	100	0.232	2.25 min	1.78
^{28}Si	92		*	
^{29}Si	5		*	
^{30}Si	3	0.1	2.62 h	none
^{55}Mn	100	13.3	2.58 h	0.85
^{63}Cu	69	4.5	12.9 h	0.51
^{65}Cu	31	2.3	5.1 min	1.04

* means that the nucleus formed is not radioactive. The neutron cross section measures the relative probability of the nucleus absorbing a neutron.

The experiment is carried out as follows. A sample of the alloy is weighed out, along with a similar sample of pure Mn metal. The two samples are irradiated in the same neutron flux for the same period of time. They are then left to stand for 20 minutes, after which the metal sample is stood for 10 minutes on a scintillation counter; the counter is fitted with a 'gate' which excludes γ-rays of energies less than 0.8 MeV. The alloy then replaces the metal for a further 10 minutes, after which the scintillation counter is run for a further 10 minutes with no sample in it. The readings of the pure metal and the alloy need a small correction to take account of the decay during the time after they were removed from the flux, and another for the background radiation; the count rate per gram of the Mn metal can then be found, and hence the mass of Mn in the alloy.

---

1 What is meant by 'qualitative and quantitative analysis of chemical elements' (para. 1)? [2]

2 Explain what is meant by the half-life of a radioisotope. [1]

3 Give the numbers of protons and neutrons in a ^{23}Na and a ^{24}Na nucleus. What changes occur in the nucleus when a ^{24}Na nucleus decays into a ^{24}Mg nucleus? [4]

4 Why is it important that the two samples in the experiment experience the same neutron flux? [1]

5 Give two reasons why the presence of a small amount of Mg does not interfere with the analysis of Mn in this experiment. [2]

6 What fraction of the original radioactivity caused by the Al will remain after $4\frac{1}{2}$ minutes standing? What fraction, roughly, therefore will remain after 18 minutes? [4]

7 Explain why the presence of Cu does not interfere with the analysis of Mn in this experiment. [2]

8 Why is the scintillation counter run for 10 minutes with no sample in it? [1]

9 Explain in your own words why the scintillation counter reading for the alloy needs 'a small correction' (para. 4). [1]

10 What further experiment could be readily performed to confirm that the γ-rays were indeed due to the Mn, and not to some unsuspected element? [2]

6 Ozonolysis

Many naturally occurring compounds such as terpenes and rubbers consist of molecules which contain $C=C$ double bonds. Many have complex structures whose analysis can be difficult. A useful technique is ozonolysis, in which an organic molecule is treated with ozone; this causes the molecule to break in two across the double bond. The products of the reaction are therefore smaller molecules, which can often be more readily identified.

Ozone has the formula O_3, and is obtained by passing oxygen, O_2, through an electric discharge. The conversion is not complete, but there is no reason to obtain the ozone pure in this case, and so 'ozonised' oxygen can be passed directly into the alkene, dissolved in an organic solvent.

The initial product of the reaction is called an 'ozonide', but these compounds can explode when purified, and so they are decomposed by the addition of water. If this is done under reducing conditions, for example in the presence of zinc and dilute acid, the products of the reaction are aldehydes and ketones:

$$\underset{R_2}{\overset{R_1}{>}}C=C\underset{R_3}{\overset{H}{<}} \xrightarrow{O_3} \underset{R_2}{\overset{R_1}{>}}C=O \quad + \quad O=C\underset{R_3}{\overset{H}{<}}$$

If oxidising conditions are used, for example by adding hydrogen peroxide, then carboxylic acids and ketones are formed:

$$\underset{R_2}{\overset{R_1}{>}}C=C\underset{R_3}{\overset{H}{<}} \xrightarrow[H_2O_2]{O_3} \underset{R_2}{\overset{R_1}{>}}C=O \quad + \quad R_3-C\overset{\diagup O}{\underset{\diagdown OH}{}}$$

If the molecule contains more than one double bond, then each of them is broken.

In general, ozonolysis of an alkene gives rise to two products, which may be separated by chromatography or distillation. However, symmetrical alkenes, regular polymers and ring compounds give rise to a single product.

Although ozonolysis gives useful information about the structure of alkenes, it does not always identify their structures completely. When a molecule contains more than one double bond, it is not always clear how the aldehydes and ketones should be put back together to form the original molecule. Even in cases where there is only one double bond, the technique does not distinguish between *cis* and *trans* isomers, which behave identically when the $C=C$ bond is broken.

An alternative to ozonolysis is the use of sodium iodate(VII) ($NaIO_4$) in the presence of a trace of potassium manganate(VII) ($KMnO_4$). It is known that potassium manganate(VII) oxidises alkenes to 1,2-diols (for example, ethene forms $HO-CH_2-CH_2-OH$); the sodium iodate(VII) is then thought to reoxidise the manganese back to manganate(VII), and break the C−C bond between the −OH groups:

$$R_1R_2C-CHR_3 \longrightarrow {R_1 \atop R_2}{>}C=O \quad + \quad {H \atop R_3}{>}C=O$$
$$| \quad |$$
$$OH\ OH$$

1 Draw the structural formula of propene, and hence identify the products which would be obtained by treating propene with ozone, followed by addition of water under reducing conditions. [3]

2 What alkene would give after ozonolysis a 1:1 mixture of propanone and methanal? [2]

3 Why are aldehydes affected if hydrogen peroxide solution is added after treatment with ozone, while ketones are unaffected? [1]

4 By considering an ozonolysis which produces ethanal as the only product, explain why ozonolysis 'does not distinguish *cis* and *trans* isomers' (para. 5). [3]

5 By choosing a suitable example, show how ring compounds can give rise to only one product on ozonolysis. [2]

6 Why is only a trace of potassium manganate(VII) needed when sodium iodate(VII) and potassium manganate(VII) are used to break C=C bonds? What role does $KMnO_4$ play here? [2]

7 What type of compound other than alkenes would give aldehydes and ketones on treatment with sodium iodate(VII) and potassium manganate(VII)? [1]

8 γ-terpinine is found in coriander oil. Its formula is:

$$CH_3-C{\overset{CH_2-CH}{\underset{CH-CH_2}{<}}}{>}C-CH{\overset{CH_3}{\underset{CH_3}{<}}}$$

Predict what substances would be formed from γ-terpinine after ozonolysis under reducing conditions. Suggest the formula of another hydrocarbon which would give the same products on ozonolysis. [4]

9 Natural rubber consists of a polymer which contains C=C bonds; on ozonolysis it gives $O=CH-CH_2-CH_2-CO-CH_3$ as the only product. Write down the formula of natural rubber. [2]

7 Arsenic

Arsenic compounds have been known for many thousands of years; some pigments containing arsenic have been found in cave paintings. In the 18th century arsenic compounds were used to colour paints and wallpapers, a practice which was not abandoned for over a hundred years, when it was realised that most arsenic compounds are highly poisonous. The 'arsenic' of detective stories is in fact arsenic(III) oxide; 0.2 g of the white solid, administered in water, is hard to detect, and fatal. Arsenic compounds still find some use as insecticides, and more recently in semiconductors.

A test for arsenic was first discovered in 1836. The substance is added to zinc and dilute acid, which reduces any arsenic present to arsine, AsH_3. This is carried along with the hydrogen evolved through a narrow heated tube. The arsine is decomposed, and a dark streak of arsenic is formed on the cooler parts of the tube. The test is highly sensitive.

Arsenic is the third member of Group V of the Periodic Table. It exists in at least two allotropic forms. The yellow form appears on rapid cooling of the vapour, and is an insulator, probably containing As_4 molecules. On standing it turns into the grey form, which has a much higher density, and conducts electricity.

Arsenic reacts directly with metals, forming metal arsenides, which are readily hydrolysed by water, forming arsine. It also forms non-metal arsenides which are covalent; $AsCl_3$ is a colourless liquid. Arsine is stable, but not appreciably basic. Arsenic(III) oxide is a white solid which dissolves a little in water forming a mildly acidic, colourless, tasteless solution; it reacts with sodium hydroxide forming sodium arsenate(III), Na_3AsO_3, and with hydrogen sulphide forming arsenic(III) sulphide. On gentle heating, arsenic(III) oxide sublimes. Arsenic forms a small number of compounds in which As has the oxidation number +V, but they are less common than the corresponding phosphorus compounds.

Arsenic is not a very common element, but it is quite widely dispersed through the earth's crust, which can give some problems to those who extract other metals commercially. Arsenic(III) oxide is found in the flue gases of roasters and smelters, and must not be vented into the air. It is separated from the flue gases, and then acts as a major source of arsenic.

1 a) What physical evidence is there that suggests that arsenic should be classified as a metal? [1]

   b) What chemical evidence is there that suggests that arsenic should be classified as a non-metal? [2]

2 Carbon and arsenic both exist in two allotropic forms; in what way does the allotropy of arsenic differ from that of carbon? [1]

3 Suggest what the formula of sodium arsenide might be, and hence write an equation for its reaction with water (para. 4). [3]

4 What properties of arsenic(III) oxide made it so convenient for poisoners? [2]

5 Would you expect antimony, the element below arsenic in Group V, to be more or less metallic than arsenic? [1]

6 What shape would you expect the arsine molecule to be, and why? In what way does arsine differ from ammonia? [3]

7 a) When arsenic(III) oxide is heated to 500 K at a pressure of 10^5 Pa, its density is about 9.6 g dm^{-3}. Calculate the number of moles of gas in 1 dm^3 under those conditions, using $PV = nRT$, and hence calculate the approximate relative molecular mass of the compound. $(1 \, dm^3 = 10^{-3} \, m^3, R = 8.31 \, J \, mol^{-1} K^{-1})$. What must its molecular formula be? (A_r: As 75, O 16) [5]

   b) Given that the arsenic atoms are arranged in a tetrahedron, and that the oxygen atoms are all chemically identical, suggest a structural formula for the gas phase molecule. [2]

8 Catalysis of Organic Reactions

In recent years much attention has been paid to transition metal compounds that can act as catalysts in organic reactions. At first the work was largely empirical, but it soon became apparent that unusual reactions could be carried out under milder conditions and in good yield. More recently the mechanisms of some of these reactions have become understood; this is very important if chemists are to control what reactions are occurring, rather than just proceed by trial and error. An example is given below.

The following stages are thought to be how a rhodium compound catalyses a particular chemical reaction.

1. $$CH_3OH + HI \longrightarrow CH_3I + H_2O$$

2. $$Rh(CO)_2I_2{}^- + CH_3I \longrightarrow Rh(CO)_2I_3(CH_3)^-$$

3. $$Rh(CO)_2I_3(CH_3)^- \longrightarrow Rh(CO)I_3(COCH_3)^-$$

4. $$Rh(CO)I_3(COCH_3)^- + CO \longrightarrow Rh(CO)_2I_3(COCH_3)^-$$

5. $$Rh(CO)_2I_3(COCH_3)^- \longrightarrow Rh(CO)_2I_2{}^- + CH_3COI$$

6. $$CH_3COI + H_2O \longrightarrow CH_3COOH + HI$$

The evidence which points to this reaction scheme includes:
i)   The reaction is zero order in CH_3OH and CO.
ii) The reaction is first order in $Rh(CO)_2I_2{}^-$, and in HI.
iii) The reactions of methanol and hydrogen iodide, and of ethanoyl iodide (CH_3COI) and water, are known to be fast under the reaction conditions. The reaction is always carried out with much more methanol than hydrogen iodide present, and the rate determining step is step 2.

1 By considering which ions and molecules are created in one step, but destroyed in another, show what net chemical change is produced by this sequence of chemical reactions. [2]

2 What does 'the reaction is zero order in CH_3OH and CO' mean (para. 3)? [1]

3 Give a rate equation for the reaction; what is the overall order of the reaction? [2]

4 In which steps is a $C-I$ bond broken? [2]

5 In which steps is a $C-I$ bond formed? [2]

6 In which step is a $C-C$ bond formed? [1]

7 In which step does the coordination number of the Rh change from 4 to 6? [1]

8 Show how the mechanism proposed, including which step is rate determining, explains the observed fact that the reaction is zero order in CO. [2]

9 The ion $Rh(CO)_2I_2^-$ is square planar. Draw two possible isomers of the ion. How many isomers would there be if the ion were tetrahedral? [3]

10 Remembering that CO is a neutral molecule, calculate the oxidation number of Rh in $Rh(CO)_2I_2^-$. [1]

11 Give an example, other than the one in this exercise, of a reaction in which a new $C-C$ bond is produced.
Use your answer to suggest a scheme whereby the product of the catalysed reaction could be formed from bromomethane. [3]

9 Clathrates and Zeolites

It has been known for some time that when quinol, 1,4-dihydroxybenzene, is crystallised from solution in the presence of argon at high pressure, the solid differs from pure quinol, and on melting gives off argon. The solid is not a compound of argon, but contains noble gas atoms trapped in holes between the quinol molecules. Three quinol molecules can be shown to form a roughly spherical cage, and many of these holes are filled by argon atoms. The solid has no fixed formula, but the minimum ratio of quinol : argon is 3 : 1. Such a solid is called a clathrate, and quinol forms one with a number of molecules. O_2, NO, Kr, Xe, CH_3OH and HCOOH all form clathrates, but C_2H_5OH and the higher alcohols, and He, do not. Clathrates are known with other compounds; water acts as the host lattice for Ar, Kr and CH_3Cl.

Another group of solids whose lattices can incorporate small molecules are the zeolites. Zeolites are aluminosilicates; their structures are based on the tetrahedral structure of SiO_2, with some of the Si atoms being replaced by Al atoms. Each substitution requires another metal cation to be incorporated to maintain electrical neutrality. Zeolites can also absorb water with little change in structure. The general formula for zeolites is $M_2O . Al_2O_3 . xSiO_2 . yH_2O$, where M is a Group I metal. They have open crystal structures, with large holes which the metal ions and water molecules occupy. These holes have radii 0.2–1.2 nm, and suitably sized ions and molecules can pass in and out relatively easily. Zeolites have therefore been used as ion-exchangers and molecular sieves. The holes differ from those in clathrates in that the ions and molecules experience stronger attractive forces from the surrounding atoms.

A zeolite can be used for water softening if it contains Na^+ ions in its holes. When hard water is passed over the zeolite, the Ca^{2+} and Mg^{2+} ions replace the Na^+ ions, so that soft water emerges. When all the Na^+ ions are used up, the zeolite may be regenerated by passing a concentrated salt solution over it. Zeolites have also been used as molecular sieves; when a mixture of branched and straight chain alkanes is passed over a zeolite, the straight chain isomers are absorbed, while the more bulky branched chain isomers are not. Small amounts of oxygen are similarly removed from argon; the larger argon atoms are unable to enter the zeolite. One of the most important applications of zeolites is their use as catalysts. In the conditions inside a zeolite cage it has been possible to convert methanol into short-chain hydrocarbons, and to crack long-chain hydrocarbons.

The front cover photograph shows a model of a zeolite; three cages are occupied by Na^+ ions and the fourth contains an alkane molecule. The photograph was kindly supplied by BP Research Ltd.

1 Give the structural formula of quinol. [1]

2 Give two pieces of evidence that suggest that clathrates are not true compounds. [2]

3 Suggest why ethanol and the higher alcohols do not form clathrates with quinol. [1]

4 Suggest why helium does not form a clathrate with quinol. [1]

5 If there is no chemical bonding between a Kr atom and a quinol molecule, suggest what weak forces might exist between them in a clathrate. [1]

6 How would you expect the enthalpy change for the melting of the argon–water clathrate to compare with that for the melting of pure ice? [1]

7 Explain in terms of oxidation numbers why the Na:Al ratio in a sodium-containing zeolite must be 1:1. [2]

8 Why is it desirable to remove calcium and magnesium ions from water? [1]

9 What are short-chain alkanes used for? Why might it be desirable to separate straight and branched chain alkanes (para. 3)? [2]

10 What is argon used for? Why might it be important to remove small amounts of oxygen from it before use (para. 3)? [2]

11 What is meant by 'to crack long-chain hydrocarbons'? Give two reasons why cracking is valuable. [3]

12 a) The absorption of water into zeolites is usually accompanied by evolution of heat; does this imply bond formation or breaking?
   b) What sort of bonds are involved in this case?
   c) Would you therefore expect CH_3OCH_3 or $CH_3CH_2CH_3$ to be absorbed more strongly by a zeolite? [3]

10 Kinetic and Thermodynamic Control

There are many examples in organic chemistry where a set of reactants may give rise to more than one set of products. The addition of hydrogen bromide to propene is such an example: the HBr could add either way across the double bond. In this case the two possible products cannot convert into each other, and the relative proportions of the products depend on the rates at which they are formed. This in turn depends on the activation energies of the reactions. The product of the reaction is said to be under kinetic control.

A second situation arises where one of two possible products converts rapidly into the other. An example of this is the hydrolysis of dichloromethane, which could give methan-1, 2-diol, but which actually forms methanal. As methan-1,2-diol reacts rapidly to form methanal and water, it does not matter which of the possible products forms faster; the more stable product is formed, and the reaction is thermodynamically controlled.

A third situation can arise where one product is formed more rapidly, but then slowly converts into the other, more stable, product. Either product may then be obtained, depending on the time taken for the reaction. An example of this is the addition of hydrogen chloride to 2-methylbuta-1,3-diene, described below.

Dissolve 5 g of 2-methylbuta-1,3-diene in an organic solvent, and place it in a flask with a gas inlet. Weigh the flask, and cool it to $-20\,°C$. Pass hydrogen chloride gas into the flask until the ratio of HCl to 2-methylbuta-1,3-diene is 1 : 2; this is determined by weighing the flask periodically. Take half the mixture, add solid calcium carbonate, and then anhydrous calcium chloride; distil off the organic product, 2-chloro-2-methylbut-3-ene. To the other half of the mixture add more HCl, and then keep it at $-20\,°C$ for 24 hours. The low temperature prevents the HCl from adding across the second double bond. The mixture is then treated as before, and 1-chloro-3-methylbut-2-ene is distilled off.

---

1 Give the structural formulae of the two possible products of the reaction of propene and hydrogen bromide. [2]

2 Explain why a reaction with a lower activation energy will occur more rapidly than one with a high activation energy (para. 1). [2]

3 Give an equation for the hydrolysis of dichloromethane by water. [1]

4 Give equations for the two possible reactions of 2-methylbuta-1,3-diene with hydrogen chloride, including the structural formulae of the organic compounds. [3]

5 What increase in mass is required (para. 4) before the first sample of the reaction mixture is removed? (A_r: H 1, C 12, Cl 35.5) [3]

6 What is the purpose of adding calcium carbonate and anhydrous calcium chloride (para. 4)? [2]

7 In this experiment (para. 4), which product is kinetically favoured, and which thermodynamically favoured? [1]

8 Give the structural formula of a product that might form if the mixture were allowed to stand at room temperature for some time (para. 4). [1]

9 Draw energy profiles for the conversion of 2-methylbuta-1,3-diene into the two products, indicating clearly the relative activation energies and enthalpy changes for the two reactions. The formation of 2-chloro-2-methylbut-3-ene is slightly exothermic. [4]

10 Suggest a simple property which could be used to show that the two products formed in this reaction are indeed different. [1]

11   Vinyl Chloride

An important bulk chemical is chloroethene, $CH_2=CHCl$, widely known as vinyl chloride. It is made on a large scale from ethene in a two-stage process. Ethene is treated with chlorine in solution, and the chlorine adds across the double bond; the product $Cl-CH_2-CH_2-Cl$ is a liquid, and acts as a convenient solvent for the reaction. The reaction is carried out at room temperature, with excess heat being removed from the reaction vessel. The product is then heated to about 800 K, where it loses HCl, forming vinyl chloride. The reaction is a free radical chain reaction, initiated by a $C-Cl$ bond breaking:

$$Cl-CH_2-CH_2-Cl \rightarrow Cl-CH_2-CH_2\cdot + Cl\cdot$$

and then propagating mainly through:

$$Cl\cdot + Cl-CH_2-CH_2-Cl \rightarrow HCl + Cl-CH_2-CHCl\cdot$$

$$Cl-CH_2-CHCl\cdot \rightarrow CH_2=CHCl + Cl\cdot$$

The vinyl chloride is distilled to purify it.

Much vinyl chloride is polymerised to give polyvinyl chloride, PVC. This is a porous, white, rigid solid, and is the second most widely used polymer. The polymerisation is carried out in water, under pressure to prevent the vinyl chloride boiling off. An organic peroxide is used as an initiator, and the polymer is separated from the mixture by centrifuging.

Vinyl chloride is also used as the starting material for the production of 1,1,1-trichloroethane, a valuable solvent. The first stage is the addition of hydrogen chloride at room temperature, forming 1,1-dichloroethane, in accordance with Markovnikov's law. This is then treated with chlorine at high temperature, when free radical substitution occurs. This reaction is not very selective, and so other polychlorinated alkanes can be formed; the desired product is separated by fractional distillation.

1 Give the name of $Cl-CH_2-CH_2-Cl$, and the structural formula of 1,1,1-trichloroethane. [2]

2 Give the structural formula of PVC. [1]

3 How are ethene and chlorine manufactured on a large scale? [2]

4 Which of the reactions described above is an elimination reaction? [1]

5 a) In the reaction in which vinyl chloride is formed (para. 1), which two radicals act as chain carriers? [2]
   b) Explain why the reaction is a chain reaction. [2]
   c) Which step in the reaction is certainly endothermic? [1]
   d) Suggest a reaction which is likely to terminate the chain. [1]

6 a) Explain what is meant by an 'organic peroxide' (para. 2). [1]
   b) How does the peroxide form free radicals? [1]
   c) Is the peroxide a catalyst in the reaction? [2]

7 What alternative product, forbidden by Markovnikov's law, is not formed when vinyl chloride is treated with hydrogen chloride? [1]

8 What inorganic product is formed in the chlorination of 1,1-dichloroethane? [1]

9 Suggest two unwanted polychlorinated alkanes which might be formed when 1,1-dichloroethane is chlorinated (para. 3) [2]

12 Manufacture of Bromine

Sea water is a useful source of several elements; its composition does not vary greatly over the surface of the earth, and the major elements dissolved in it are:

Element	Abundance/$g\,kg^{-1}$
Cl	19
Na	11
Mg	1.3
S	0.9
Ca	0.4
K	0.4
Br	0.065

Although bromine is a relatively small constituent of sea water, it is worthwhile to extract it commercially.

The bromine in sea water is present as bromide ions. The extraction begins with sea water being mixed with acid to bring the pH to about 3.5; chlorine is then added, converting the bromide ions to bromine. The pH is controlled to prevent the chlorine being hydrolysed to HCl and HClO, and bromine to HBr and HBrO. The mixture is then passed down a blowing out tower, in which a counter current of air removes the bromine from the water. The vapours at the top of the blowing out tower are treated with a mixture of sulphur dioxide and water, which converts the bromine to hydrogen bromide. Any chlorine present is similarly converted to hydrogen chloride. The liquid formed typically contains about 13% HBr, 9% H_2SO_4 and 1% HCl. It is then treated in a steaming tower with a counter current of chlorine and steam, which reforms bromine vapour. The hot vapour from the tower is condensed to form crude bromine, which is then distilled, condensed and dried. Overall, 20 kg of sea water produces just 1 g of bromine. The acid solution left from the steaming out is used to acidify more incoming sea water.

The major use of bromine is in the formation of 1,2-dibromoethane. This is added to petrol to prevent a build up of lead in the engine. When leaded petrol is burnt, small amounts of PbO are formed, and these would accumulate in the engine, causing performance to deteriorate. 1,2-dibromoethane converts the PbO into $PbBr_2$, which is volatile at engine temperature; the lead then passes out with the exhaust gases. Bromine is also used to make bromomethane and dibromomethane, both of which are used as fumigants for crops such as tobacco and strawberries.

1 It is often said that sea water is mostly a solution of sodium chloride in water; is that compatible with the data given opposite? [1]

2 Give an equation for the reaction of bromide ions with chlorine. [1]

3 Write a balanced equation for the hydrolysis of chlorine. The pH of sea water is adjusted to 3.5 to prevent the chlorine being hydrolysed; use Le Chatelier's principle to show whether this value is the maximum or minimum possible. [2]

4 How is the chlorine needed in the extraction of bromine manufactured commercially? [2]

5 Give an ionic equation for the reaction of sulphur dioxide and bromine in the presence of water, and identify the change in oxidation state of the sulphur. [3]

6 Use the data given in the passage to calculate the overall efficiency of the extraction of bromine from sea water. [2]

7 How, and under what conditions, might 1,2-dibromoethane be formed from bromine? [2]

8 Would you expect demand for bromine to increase or decrease in future? Explain your reasoning. [2]

9 How can bromomethane and dibromomethane be made from bromine? [3]

10 What would be the major problem to be overcome in designing the plant for the extraction of bromine? [2]

13   A Pseudo-Halogen

The group $-C\equiv N$ is quite stable, and is found in a variety of chemical compounds. Because it has valency 1, and can be found in ionic and covalent compounds, it resembles the halogens, and is referred to as a pseudo-halogen.

Hydrogen cyanide is a volatile liquid which smells of almonds; it is extremely poisonous. It is soluble in water, and a weak acid, and so it is formed when metal cyanides are treated with mineral acids. It can be manufactured by the reaction of methane with ammonia, an endothermic reaction which takes place at 1200 °C on a Pt catalyst. Hydrogen is also formed.

Sodium cyanide is a white crystalline salt, with the same crystal structure as sodium chloride. The cyanide ion is isoelectronic with carbon monoxide, and can act as a base, a nucleophile and a ligand. The cyanide ion is used as a nucleophile in organic chemistry, for introducing the $-CN$ group; it can then be hydrolysed in acid to give a $-CONH_2$ group, and then a $-COOH$ group, or it can be reduced with hydrogen to form a primary amine. The cyanide ion is used as a ligand in the formation of 'iron blues', in which Fe ions are surrounded by six cyanide ions; these are used in printing inks. Cyanide ions are also used in the extraction of silver metal:

$$4Ag + 8CN^- + 2H_2O + O_2 \longrightarrow 4Ag(CN)_2^- + 4OH^-$$

Cyanogen, $NC-CN$, is the parent pseudo-halogen, and can be formed from cyanides in ways similar to those used for the halogens themselves. It is an important intermediate in the formation of fertilisers such as $H_2NCO-CONH_2$.

Hydrogen cyanide and many of its compounds are very toxic, and disposal of them is a major practical problem. Cyanides are rendered quite safe by complexing them with Fe^{2+} ions, which modify the properties of the cyanide ions enough to prevent HCN being formed readily, but they are better treated by hydrolysis followed by oxidation, forming ammonia, carbon dioxide and water.

1 By choosing two suitable examples, show how halogens have valency 1 and are found in ionic and covalent compounds (para. 1). [2]

2 Show (by a dot-and-cross diagram or otherwise) the arrangement of the valence electrons in a cyanide ion; what feature of the structure means that it can act as a base or nucleophile? [3]

3 CN^- and CO are isoelectronic; explain the meaning of this sentence, and suggest another molecule which is isoelectronic with them. [2]

4 Give a balanced equation for the formation of HCN from methane and ammonia. [2]

5 Suggest two reasons why a very high temperature is used in the formation of HCN. [2]

6 Give a balanced equation for the hydrolysis of HCN in acid (include both steps). [2]

7 Suggest what might be formed if NaCN were treated with (a) chlorine, (b) silver nitrate solution. [2]

8 Sodium cyanide is used in many laboratories for organic synthesis; suggest two safety precautions which would be sensible in those labs. [2]

9 Suggest in outline schemes whereby the following could be made from HCN or NaCN: (a) CH_3NH_2 (b) $Fe(CN)_6^{4-}$ (c) $H_2NCO-CONH_2$. [3]

14 Fire and Fire Retardants

The damage caused by fire continues to increase steadily. Although fire causes damage to property, it is also a major cause of death. The most obvious hazard with fire is the structural damage of the fire itself, although it is now recognised that equally important is the asphyxation caused by fumes from the fire. Where there is technical equipment, damage caused by acidic gases in smoke can be severe.

The ease of combustion of organic material is measured by the limiting oxygen index. A sample is clamped and ignited; it is surrounded by an oxygen/nitrogen mixture whose composition can be varied. The limiting oxygen index is the percentage of oxygen which will just sustain a flame. Some typical values are:

Perspex	17.3	Polythene	17.4
Cotton	18.4	Nylon fibre	20.1
Polyester fibre	20.6	Wool	25.2
PVC	37.1	Teflon	95

Substances with limiting oxygen indices of more than 21 are classified as self-extinguishing.

All polymers will burn if the temperature is high enough, but there are several ways of reducing the likelihood of a flame causing a major fire, or of increasing the ignition temperature. Aluminium hydroxide is used widely as an additive in organic products; on heating it gives off water, which dilutes the flammable gases given off when a polymer burns, and so reduces the temperature of the flame. This in turn reduces the rate of heat transfer to the polymer itself, and makes it less likely that the fire will be self-sustaining.

Hydrogen chlorine is also well known as a flame inhibitor. The chemical reactions in flames involve free radicals, with $OH\cdot$ and $H\cdot$ radicals being particularly reactive. Hydrogen chloride inhibits flames by reacting with $OH\cdot$ and $H\cdot$ radicals, forming $Cl\cdot$ radicals, which are much less reactive,

e.g.
$$OH\cdot + HCl \longrightarrow H_2O + Cl\cdot$$

The HCl molecule is then regenerated by the $Cl\cdot$ reacting with, for example, an alkane molecule in the flame:

$$Cl\cdot + CH_4 \longrightarrow HCl + CH_3\cdot$$

Most halogen-based fire retardants are organic compounds which produce HCl or HBr on heating.

An alternative approach to flame retarding is to modify the properties of the solid polymer by incorporating additives. An endothermic decomposition (such as that of aluminium hydroxide) will reduce the temperature of a polymer if it catches fire, and so reduce the

COMPREHENSION

production of flammable gases. Physical protection can also be produced by an additive which decomposes to form a charred solid matrix on the polymer surface.

1   Give three conditions necessary for combustion to occur.   [2]

2   Which burns more readily, cotton or PVC? Explain your answer.   [2]

3   Why are substances with a limiting oxygen index above 21 classified as self-extinguishing?   [2]

4   a) The limiting oxygen index is one measure of how readily a substance will burn; suggest two limitations of it as a measure of the potential danger of a fabric.   [2]
    b) The heat of combustion would be another useful property with which to discuss the fire risks of fabrics; in what units would it be measured?   [1]

5   Give an equation for the thermal decomposition of aluminium hydroxide, and explain two ways in which it acts as a fire retardant.   [3]

6   a) Draw a diagram to show the arrangement of electrons in an H· and an OH· radical.   [2]
    b) One of the steps in combustion which is thought to be inhibited by HCl is:

$$OH\cdot + CO \rightarrow CO_2 + H\cdot$$

Estimate the enthalpy change for this reaction, given the bond energies H−O 464, C≡O 1077, C=O 805 kJ mol^{-1}.   [2]

7   What problems may arise through the use of organic halogen compounds as fire retardants?   [2]

8   A student suggests that the idea that HCl can help to reduce fires is wrong, as Hess's Law states that the total energy given out in a chemical reaction is independent of the route taken; details of the mechanism do not affect the total amount of heat produced. Explain why this argument is wrong.   [2]

15 Differential Thermal Analysis

Differential Thermal Analysis, DTA, is a valuable technique for studying physical and chemical changes that take place in the solid state. DTA consists of measuring temperature differences between a sample being investigated and a reference sample, both of which are heated steadily under identical conditions. While no change takes place in the sample, the temperature of the two samples rises together, so there is no temperature difference. If an endothermic change, such as melting, takes place in the sample, then its temperature falls below that of the reference, and stays below until the change is completed, when the temperature difference returns to zero. If an exothermic reaction occurs, then the temperature of the sample rises above that of the reference until the change is complete. DTA therefore identifies physical and chemical changes, and the temperatures at which they occur. Measuring the areas under the peaks can give values for enthalpy changes.

DTA has been used to investigate melting points, allotropic transitions, loss of water of crystallisation, ring closures and thermal decompositions. The temperatures of the sample and the reference are measured by thermocouples, and the difference in voltage is then measured and displayed. The temperature of the heating block should rise steadily with time; the DTA plot is then one of temperature change (ΔT) against temperature (T). The samples may be very small, with as little as 10 mg being used; aluminium oxide is often used as the reference material.

Two examples are given below. Figure 1 shows the DTA plot for hydrated calcium ethanedioate, $CaC_2O_4 \cdot H_2O$ under two different conditions: with the sample surrounded by (a) air, and (b) He. The plot shows endothermic changes below the base line. The interpretation is thought to be: at 200 °C the salt loses water of crystallisation, at 450 °C it loses CO, giving $CaCO_3$, and at 850 °C the $CaCO_3$ loses CO_2. In He all the changes are endothermic, but when oxygen is present the CO is oxidised to CO_2, making the second decomposition exothermic.

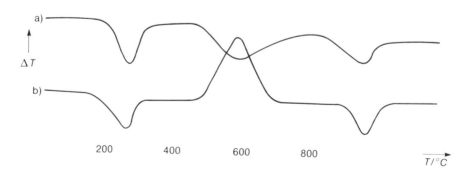

Figure 1

Figure 2 shows the DTA plot for $KClO_3$ (a) with and (b) without the addition of 10% MnO_2. In each case the $KClO_3$ melts at 300 °C, and without the MnO_2 there are two further processes, the disproportionation

$$4KClO_3 \longrightarrow 3KClO_4 + KCl$$

and the decomposition

$$KClO_4 \longrightarrow KCl + 2O_2$$

When the MnO_2 is added there is only one process,

$$2KClO_3 \longrightarrow 2KCl + 3O_2$$

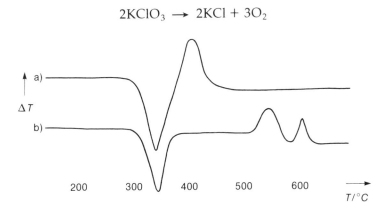

Figure 2

---

1 Give two properties that a substance such as aluminium oxide must have to make it a suitable choice as a reference material. [2]

2 Why might a thermocouple be more suitable for measuring temperature changes than an ordinary mercury thermometer? [2]

3 Explain why in interpreting a peak in a DTA plot the temperature of a transition is taken as the temperature at the *lowest* point on the peak, rather than at the centre of the peak. [2]

4 Suggest one reason why DTA is a suitable technique for studying explosive substances such as ammonium nitrate. [2]

5 What is meant by the term 'allotropic transitions' (para. 2)? Give an example of such a transition. [2]

6 Is the dehydration of calcium ethanedioate exothermic or endothermic? [1]

COMPREHENSION

7 Suppose you were investigating the DTA plot of hydrated calcium ethanedioate for the first time. What further DTA experiment would you do to confirm your tentative interpretation of the peak at 850 °C? What would you look for in that experiment? [2]

8 Given the following heats of formation: $CaC_2O_4 - 1361$, $CaCO_3 - 1207$, $CO - 111$, $CO_2 - 394 \, kJ \, mol^{-1}$, construct cycles to predict the enthalpy changes for the decomposition of CaC_2O_4 in air and in He. Are your results compatible with the DTA plots? [4]

9 What role does MnO_2 play in the decomposition of $KClO_3$? [1]

10 If less MnO_2 is used in the mixture of $KClO_3/MnO_2$, the decomposition still occurs at the same temperature, but more slowly. Sketch on the same graph the DTA plots for mixtures containing 10% and 5% MnO_2, assuming the same rate of heating in each case. [2]

16 The Alkali Metals

The reactions of metals with air are not always as simple as might be imagined. M and X are two elements in Group I of the Periodic Table. Each is burnt separately in dry air in a crucible, producing white powders Y and Q respectively, both of which produce colours in a flame test.

When water is added to Y it forms a colourless solution R, which turns a red litmus paper blue. On warming, the solution R gives off an alkaline gas Z, which forms white fumes when mixed with hydrogen chloride gas. The solution R gives no precipitate when sodium carbonate solution is added to it. The solid Y contains two substances T and A; T can also be obtained by adding M to water, and then heating the solution to dryness.

When water is added to Q, and the mixture warmed in the presence of a little manganese(IV) oxide catalyst, oxygen is evolved, and the resulting solution D turns a red litmus paper blue.

---

1 Give the formula of, and charge on, the negative ion which must be present in the solution R. [1]

2 What negative ion must T contain? Give equations for the reactions when M is added to water, and the solution then heated to dryness. [3]

3 Suggest the identity of the gas Z, and give an equation for its reaction with hydrogen chloride. [2]

4 What element other than M must be present in A? Hence give the formula of A. [2]

5 Give one piece of chemical evidence that the element M cannot be a group II metal, and explain your reasoning. [2]

6 Q contains the ion O_2^{2-}; write an equation for the reaction of Q with water. [2]

7 Suggest a simple method, based on the reactions above, of determining the relative atomic mass of M. [3]

8 Describe briefly how you would conduct a flame test on Y. [2]

9 How could you obtain a pure dry sample of the chloride of X, starting from solution D? [3]

---

17 Densities of Alkali Metal Halides

The densities of the alkali metal halides are given in the table; at first sight they appear to vary irregularly, with no simple pattern emerging either as the alkali metal or the halogen increases in size. The pattern becomes clear, however, if the densities are used to calculate molar volumes (the volumes occupied by one mole of the substances). The molar volume is given by the relative molecular mass divided by the density, and molar volumes show much more regularity.

Densities of halides/$g\,cm^{-3}$

	F	Cl	Br	I
Li	2.64	2.07	3.46	3.49
Na	2.56	2.17	3.20	3.67
K	2.48	1.98	2.75	3.13
Cs	4.11	3.99	4.44	4.51

The molar volume can be used to calculate the distances between the nuclei of touching ions in the crystal lattice. For a substance with an NaCl lattice (fig. 1), the unit cell contains 4 cations and 4 anions, and has a volume of $8d^3$, where d is the distance between the nuclei of adjacent ions. It therefore follows that $4L$ cations and anions would occupy a volume of $8Ld^3$, where L is the Avogadro constant, and so the molar volume is $2Ld^3$. Some alkali metal halides adopt the CsCl lattice (fig. 2); in this case it can be shown that the molar volume is $8Ld^3/3\sqrt{3}$.

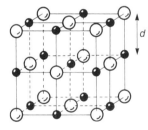

Figure 1

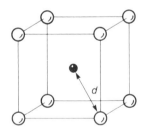

Figure 2

In the ionic model solids are seen as consisting of a regular array of ions; these are hard and incompressible, each with a definite radius. This model can be tested using the data on densities. The distance d is equal to $(r_+ + r_-)$, the sum of the radii of the cation and anion. It is not possible to measure the radius of each ion individually in this way, but as soon as the radius of one ion is fixed, then all the others may be calculated. If the ionic model holds, the values of the radius of an ion obtained from a series of its compounds should all be the same.

1   Calculate the molar volumes of the halides of Na to show that they change steadily from F to I. (A_r: Na 23.0, F 19.0, Cl 35.5, Br 79.9, I 126.9)   [2]

2   What is the coordination number of each ion in the NaCl lattice, and in the CsCl lattice?   [2]

3   a) The chlorides and iodides of sodium and potassium all have the NaCl structure. Calculate, using the formula in para. 2, the distance d between adjacent nuclei for each compound, expressing your answer in nm. (A_r: K 39.1; $L = 6.02 \times 10^{23} \, mol^{-1}$)   [4]
    b) If the radius of a Cl^- ion is 0.181 nm, calculate the radius of a Na^+ ion, and of a K^+ ion.   [2]
    c) Hence calculate two separate values for the radius of an I^- ion; do they agree with each other?   [2]

4   a) CsCl has a different structure from NaCl; use the formula in para. 2 to calculate the distance between adjacent nuclei. (A_r: Cs 132.9)   [2]
    b) Hence calculate the radius of the Cs^+ ion.   [1]
    c) The radius of the F^- ion is 0.133 nm; calculate d for CsF, and hence predict the molar volume for CsF assuming (a) that it has the NaCl structure and (b) that it has the CsCl structure.   [2]
    d) Now use the density and relative molecular mass of CsF to calculate the molar volume, and hence decide which structure is more likely to be correct.   [2]

5   Give one other method of determining the value of d for CsF.   [1]

18 Chloride Titrations

The following passage gives instructions on how the amount of chloride ion in a sample may be measured by titration.

'The sample is weighed into a stoppered flask, and distilled water is added. After any reaction is complete, the solution is treated with powdered calcium carbonate. A few drops of potassium chromate solution are added to act as indicator, and the solution is titrated with standard silver nitrate solution. At first a white precipitate is formed, and the end point is reached when the precipitate acquires a permanent red coloration.'

The following table gives the colours and solubilities of some relevant compounds.

Compound	Appearance	Solubility in water/$mol\,dm^{-3}$	Solubility in acid/$mol\,dm^{-3}$
$AgNO_3$	Colourless soln	soluble	soluble
$AgCl$	White ppt	10^{-5}	10^{-5}
Ag_2CrO_4	Red ppt	10^{-3}	soluble
K_2CrO_4	Yellow soln	soluble	soluble
KCl	Colourless soln	soluble	soluble
AgI	Cream ppt	10^{-8}	10^{-8}
KI	Colourless soln	soluble	soluble

---

1 Explain in your own words how this titration works, paying particular attention to what happens at the end point. [4]

2 List the ions which are present in significant quantities in solution when a sample of potassium chloride is titrated in this way:
a) before any silver nitrate is added,
b) when some silver nitrate has been added, but before the end point,
c) at the end point. [6]

3 Why was calcium carbonate added at the beginning of the reaction? [2]

4 How could you tell enough calcium carbonate had been added? [1]

5 Using the data above, do you think that this method could be used to measure iodide concentrations? Explain your answer. [2]

In an experiment with a chloride L, 0.107 g of L were weighed out, and found to react with 20.0 cm^3 of 0.1M silver nitrate solution.

6 How many moles of silver nitrate were used in the experiment? [1]

7   How many moles of chloride ion were there in the original sample?   [1]

8   If L has a formula of the type $X^{+6}Cl^{-6}$, calculate the molar mass of L.   [1]

9   If L has a formula of the type $X^{2+}Cl^-{}_2$, calculate the molar mass of L; hence give a general statement relating the molar mass of L and the valency of X.   [2]

19 Group IV Halides

The elements of Group IV have outer shells which are exactly half full; this gives rise to several possibilities in their chemical bonding. When they react with non-metals, such as the halogens, they could form compounds of the general formula MX_4, either by sharing their valence electrons, forming four single bonds and thereby completing their outermost shell, or by forming 4+ ions, giving away their valence electrons to the halogens. The situation is further complicated by the existence of several compounds in which the Group IV atoms show valency two; whether these are ionic or covalent, they cannot have completely filled outermost shells.

Boiling points provide a simple guideline to the nature of chemical bonding. Those of the Group IV halides are given below.

	Boiling points/K		
X =	F	Cl	Br
CX_4	145	350	462
SiX_4	187	331	427
GeX_4	236 (sub)	357	459
SnX_4	978 (sub)	387	475
PbX_4	—	378 (exp)	—
SnX_2	—	925	893
PbX_2	1563	1223	1189

(sub) means sublimes, and (exp) means that it explodes below its boiling point. Use these data to answer the questions below.

1 Describe briefly the trend in boiling points of the tetrahalides of carbon as Group VII is descended. [1]

2 Would your answer to question 1 also summarise the data for the other Group IV elements (excluding Pb)? [2]

3 How do the boiling points of the tetrahalides vary as the Group IV element is varied? [3]

4 What shape would you expect a molecule of CCl_4 to be, assuming the compound to be covalent? [1]

5 What forces are there between covalent molecules such as CCl_4? How do these forces vary with the size of the Group VII atom, and why? [3]

6 Are the boiling points given consistent with the idea that the Group IV tetrahalides are covalent? [2]

7 How do the boiling points of the dihalides of Sn and Pb vary with the size of the Group VII atoms, and with the size of the Group IV atoms? [2]

8 a) Do you think these compounds are predominantly ionic or covalent? Explain your answer.
    b) In the light of your answer explain the variation in boiling points as the Group VII element changes. [3]

9 Another method of deciding whether compounds are ionic or covalent would be to dissolve them in water, and measure the conductivity of the solution. Why might this method not work for (a) $SiCl_4$ and (b) PbI_2? [2]

10 One of the problems in using boiling points to measure intermolecular forces is that boiling points vary with pressure; under different conditions the boiling points would all change. Suggest another physical measurement which might give a more direct measure of the forces between ions and molecules. [1]

20 Oxyacids

Most non-metals form oxyacids, that is acids which contain the non-metal, oxygen and hydrogen. The acidities of these compounds vary greatly; some are strongly acidic, while others are only weakly so. Many have more than one replaceable hydrogen atom, and so have two or three successive dissociations.

The strength of a weak acid may be measured by its dissociation constant K_a. In order to keep the numbers in a convenient range, it is common to use the value of pK_a, where $pK_a = -\log_{10} K_a$, just as hydrogen ion concentrations are often expressed as pH values. Where an acid can lose successive hydrogen atoms, each ionisation has its own dissociation constant, expressed as pK_1, pK_2, etc.

The strengths of some common acids are given in the table below.

Acid	pK_1	pK_2	pK_3
HCl	large, negative		
H_2SO_4	−3(?)	1.5	
HNO_3	−1.4		
HIO_3	0.8		
H_2SO_3	1.9	7.2	
$HClO_2$	2.0		
H_3PO_4	2.2	7.2	12.3
H_2TeO_3	2.7	8.0	
HNO_2	3.3		
HClO	7.2		
HBrO	8.7		
HIO	11.0		

These values have been interpreted in terms of the structural formulae of the acids. Many oxyacids have structural formulae of the type $XO_p(OH)_q$, where the non-metal forms q bonds to −OH groups, and a further p double bonds to oxygen atoms. The acidity can then be related to the values of p and q. Once the effects of changing p and q are understood, they can be used to make predictions about other oxyacids, and hence to test whether their structural formulae are of the type above.

---

1 Explain what is meant by the dissociation constant of a weak acid. If a dissociation constant has the value $2 \times 10^{-6}\,moldm^{-3}$, what is the value of pK_a? [2]

2 Which is the stronger acid, HNO_2 or HClO? [1]

3 A solution of HClO is taken, and sodium hydroxide is added until the pH reaches 7.2. Using your answer to question 1, what can you say about the relative concentrations of HClO molecules and ClO^- ions then present in solution? [2]

4 What effect does it have on the acidity of an oxyacid if the non-metal is replaced by a larger atom in the same group of the Periodic Table? Suggest why this might be. [2]

5 Give the values of pK_1 and pK_2 for two acids of your choice, and suggest *approximately* what the difference between them is for each acid. What approximate relationship would this imply between K_1 and K_2? [3]

6 By giving data for a suitable range of acids, show what the effect is of increasing the value of p, while keeping X and q constant.
Can you suggest an *approximate* relationship between pK_1 values for $XO_p(OH)_q$ and $XO_{p+1}(OH)_q$? [3]

7 Draw the structural formulae of methanoic acid and carbonic acid (H_2CO_3). If pK_a for methanoic acid is 3.8 and for carbonic acid is 3.6, what does this suggest about the effect on pK_a of replacing an $-OH$ group with an $-H$ atom? [3]

8 Phosphorus forms either three or five covalent bonds in its compounds; it forms three oxyacids, H_3PO_2, H_3PO_3 and H_3PO_4. The structure of H_3PO_4 is

```
              O
              ||
        H-O-P-O-H
              |
              O-H
```

while that of H_3PO_3 could be

```
         O
         ||
   H-O-P-O-H        or     HO-P-OH      , and that
         |                      |
         H                     OH
```

of H_3PO_2 could be

```
         O
         ||
   H-O-P-H                   H-O-P-H
         |         or              |
         H                        O-H
```

The values of pK_1 for the acids are 1.8 for H_3PO_3 and 2.0 for H_3PO_2. By considering these values, and your answers to questions 6 and 7, explain which two structures seem more likely, explaining your reasoning. [4]

21 The Investigation of a Precipitate

Many metal compounds give precipitates of metal hydroxides when treated with sodium hydroxide solution, and these reactions can be used to identify metal ions. However, the reactions are not always as simple as they appear; the reaction of copper(II) sulphate, $CuSO_4$, with sodium hydroxide, illustrates the point.

If a solution of 0.2M sodium hydroxide is slowly added to $10\,cm^3$ of 0.2M copper(II) sulphate solution, a light green precipitate X forms. If the reaction is followed with a pH meter, or by measuring the conductivity of the solution, an end point is found when $15.0\,cm^3$ of sodium hydroxide have been added. At this point the pH changes sharply for the addition of a very small amount of alkali, and the gradient of the conductivity–volume graph changes sharply from one constant value to another.

At the end point the precipitate X is filtered off and washed carefully, leaving a colourless filtrate Y. The precipitate is completely dissolved in dilute hydrochloric acid, and the solution made up to $50\,cm^3$ with water. $25\,cm^3$ of this solution is then treated with excess KI solution, which forms iodine; the iodine is then titrated with 0.1M sodium thiosulphate solution, using starch as indicator.

$$2Cu^{2+} + 4I^- \longrightarrow 2CuI + I_2$$

$$I_2 + 2S_2O_3^{2-} \longrightarrow 2I^- + S_4O_6^{2-}$$

$10.0\,cm^3$ of sodium thiosulphate are required.

The remaining $25\,cm^3$ of the solution in dilute HCl are treated with excess barium chloride solution, and warmed to coagulate the precipitate of $BaSO_4$. The precipitate is filtered, washed until no more chloride ion is found in the washings, and then dried and weighed. The mass of barium sulphate is found to be $0.058\,g$.

---

1 Calculate the number of moles of copper(II) sulphate taken in this experiment, and the number of moles of NaOH added at the end point. [2]

2 Calculate the number of moles of thiosulphate used in the titration, and hence the *total* number of moles of copper in the original precipitate X. [2]

3 Calculate the number of moles of barium sulphate precipitated, and hence calculate the *total* number of moles of sulphate in the original precipitate X. (A_r: Ba 137.3, S 32, O 16) [2]

4 From your answers to questions 1 to 3, suggest the formula of the precipitate X. [2]

5 What does the filtrate Y consist of? How could you show in a simple experiment that it contains no Cu^{2+} ions? [2]

6 What would you *see* at the end of the thiosulphate titration? [1]

7 In para. 4 the barium sulphate is washed until no more chloride is found in the washings. Explain how you would do this, and why it is important to do it. [3]

8 Which of the following graphs might represent the variation in pH during the addition of sodium hydroxide to copper(II) sulphate? Explain your reasoning. [2]

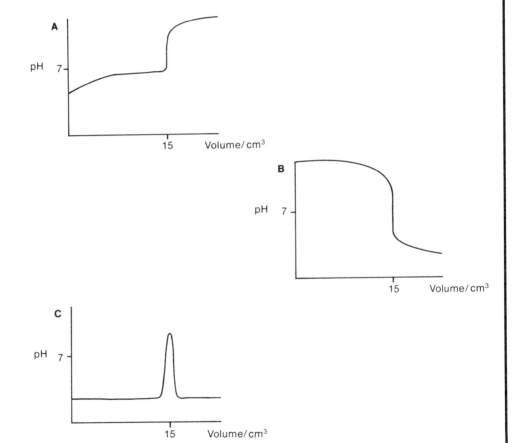

9 Which of the following graphs might represent the variation of conductivity during the addition of sodium hydroxide to copper(II) sulphate? Explain your reasoning. [2]

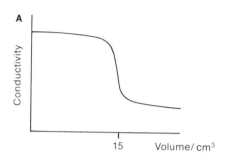

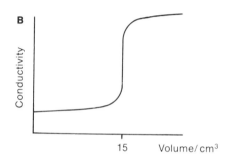

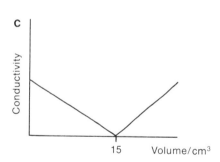

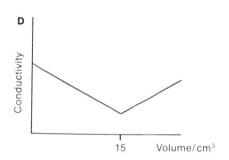

10 It is planned to repeat the experiment using copper(II) chloride solution to see whether a similar precipitate is formed. Give two alterations that would be needed in the procedure for the analysis of a precipitate. [2]

22 Covalent Bonding

In this question you will need the following data:

	H–H	Cl–Cl	O–O	N–N	C–C	H–O	H–N	C–H
Bond length/nm	0.074	0.199	0.148	0.145	0.154	0.096	0.101	0.108
Bond strength/kJ mol^{-1}	436	243	144	158	347	464	391	413

Electronegativities:　H 2.1　O 3.5　N 3.0　C 2.5　Cl 3.0

1   Define what is meant by the bond energy of a diatomic molecule such as Cl_2. [1]

2   Define what is meant by the bond enthalpy term for the C–H bond in methane. [1]

3   Why can the same definition not be used to find the bond enthalpy term for the C–C bond in ethane? How is the value of that bond enthalpy term found? [2]

4   If atoms are seen as hard spheres, then it should be true that $2r(X-Y) = r(X-X) + r(Y-Y)$, where r is the bond length. By drawing up a table for the elements C, N and O forming bonds with H, investigate whether the relationship holds. [3]

5   It has also been suggested that differences in electronegativities could cause deviations from this hard sphere model. Do your data support this idea? [2]

6   Predict the bond length of H–Cl. [2]

7   Investigate similarly the idea that bond energies are related so that $2E(X-Y) = E(X-X) + E(Y-Y)$. [3]

8   Do deviations from the equation in question 7 depend on the difference in electronegativity between X and Y? [1]

9   Given the following information: N≡N bond energy 945 kJ mol^{-1}, $\Delta H_f(NCl_3)$ +230 kJ mol^{-1}, construct a cycle to calculate the bond energy of an N–Cl bond. [3]

10   Compare your answer with the N–N and Cl–Cl bond energies, and comment. [2]

23 A Redox Reaction

Potassium manganate(VII), $KMnO_4$, is a strong oxidising agent in acid solution; it is widely used in titrations because of its strong colour change from purple to colourless when it reacts. One of the substances with which it reacts is the nitrate(III) ion, NO_2^-; this experiment is designed to identify the products of that reaction.

A solution of potassium manganate(VII) needs to be standardised before use, as it decomposes slowly on standing. This can be done with a solution of 0.1M iron(II) ammonium sulphate, $(NH_4)_2Fe(SO_4)_2$, which can be made up accurately. Fill a burette with $KMnO_4$ solution. Pipette $20\,cm^3$ of the iron(II) ammonium sulphate solution into a $300\,cm^3$ conical flask, add about $20\,cm^3$ of 1M sulphuric acid, and titrate with the potassium manganate(VII). The end point is reached when the solution just acquires a permanent pink colour. The MnO_4^- ions are reduced to Mn^{2+}, and the Fe^{2+} ions are oxidised to Fe^{3+}.

A burette is now filled with 0.1M $NaNO_2$ solution. $20\,cm^3$ of the potassium manganate(VII) solution is pipetted into a $300\,cm^3$ conical flask, $20\,cm^3$ of 1M sulphuric acid added, and the mixture warmed to about $40\,^\circ C$. The $NaNO_2$ solution is then added until the solution just turns colourless.

In an experiment the volumes used in the two titrations were $16.0\,cm^3$ in the first, and $12.5\,cm^3$ in the second. These values give information on the change of oxidation number of the nitrogen when NO_2^- is oxidised by MnO_4^-.

---

1 What is the oxidation number of Mn in MnO_4^-? [1]

2 Give an ionic equation for the reaction of MnO_4^- with Fe^{2+} in acid solution. [2]

3 What role does the sulphuric acid play in this reaction? [1]

4 What gives rise to the colour at the end point of the titration of the potassium manganate(VII) and the iron(II) ammonium sulphate? [1]

5 Calculate the number of moles of Fe^{2+} taken in the titration, and hence the number of moles of MnO_4^- needed to react with it.
What is the concentration of the MnO_4^- solution? [3]

6 a) Calculate the number of moles of MnO_4^- taken in the titration with $NaNO_2$. Hence calculate the number of moles of electrons consumed by the MnO_4^-. [2]
    b) Calculate the number of moles of NO_2^- ions taken, and hence calculate the number of electrons supplied by one NO_2^- ion. [2]

7 a) What is the oxidation number of nitrogen in NO_2^-? [1]

  b) Hence calculate the new oxidation number of nitrogen after the reaction with MnO_4^-. [1]

  c) Now suggest a way to complete the following equation:

$$NO_2^- + H_2O \longrightarrow \ldots H^+ + \ldots + \ldots e^-$$ [2]

  d) Is this the only possible answer, or merely a strong possibility? Explain your answer. [1]

8 What effect would it have on the volume of $NaNO_2$ used in the titration if:

  a) an old, unstandardised solution of $KMnO_4$ were used in the second titration,

  b) $25 \, cm^3$ of sulphuric acid were used by mistake. [2]

9 Suggest why a small error of, say, 3% in the titration reading might not be very important in this experiment. [1]

24 Hydrolysis of Methyl Methanoate

Methyl methanoate is hydrolysed in aqueous solution at room temperature, if a strong acid such as hydrochloric acid is added as catalyst. The products are methanoic acid and methanol. The extent of the reaction can be followed by titrating samples of the mixture with standard sodium hydroxide solution at regular time intervals.

The procedure is as follows. Take $200\,cm^3$ of 0.5M HCl in a $350\,cm^3$ conical flask, and add about $12\,cm^3$ of methyl methanoate. Cork the flask and shake it. As soon as possible withdraw $2\,cm^3$ of the mixture, and run it into $100\,cm^3$ of water in a second conical flask; add a few drops of phenolphthalein, and titrate with 0.1M NaOH. Repeat the withdrawal and titration every twenty minutes. Finally allow the mixture to stand for a very long time, and determine the titration volume when the reaction is completed.

A set of results obtained in this way is:

Time/min	0	20	40	60	80	final
Titre/cm^3 NaOH	10.0	16.6	21.0	24.0	26.0	30.0

1 What is the role of the phenolphthalein in the titration? [1]

2 Why is the $2\,cm^3$ sample added to $100\,cm^3$ of water? [1]

3 Why is the first titration value not zero? [1]

4 a) How many moles of NaOH react with the methanoic acid in the final titration, when the hydrolysis is completed? [2]
   b) Hence calculate the number of moles of methyl methanoate used in this experiment, and its exact volume (for methyl methanoate, density = $0.974\,g\,cm^{-3}$, $M_r = 60$). [2]

5 Calculate the number of moles of methanoic acid formed after 20 minutes, and hence the number of moles of methyl methanoate left unreacted. [2]

6 Repeat the calculation in question 5 for the other titrations. [3]

7 Plot a graph of the number of moles of methyl methanoate left against time, and hence estimate the time taken for the amount of methyl methanoate to decrease to (a) half and (b) quarter of its original value. [3]

8 What can you conclude from your answer to question 7? Explain your answer. [2]

9 On your graph sketch the line that would be obtained if the experiment were repeated using $200 \, cm^3$ of 0.25M HCl, and the same amount of methyl methanoate. [2]

10 Suggest an alternative method of finding the final titration value, that would not involve waiting for a very long time to elapse. [1]

25 Mass Spectra

Figure 1 shows the mass spectrum of a fluorocarbon A.

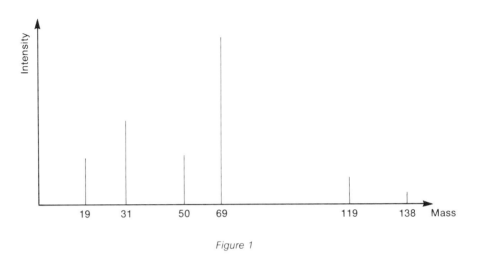

Figure 1

Compound B has the molecular formula C_2F_6Hg; its mass spectrum is shown in fig. 2. Hg has two isotopes, of similar abundance, of masses 200 and 202.

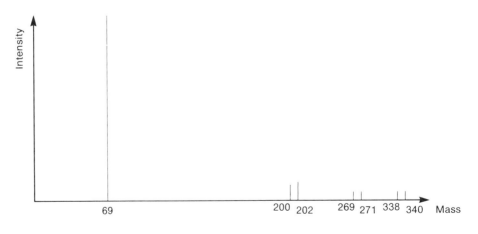

Figure 2

DATA ANALYSIS

1 What is the relative molecular mass of A? [1]

2 Suggest a molecular formula for A (A_r: C 12, F 19). [2]

3 Identify the formula of, and charges on, the ions giving rise to the peaks at mass 138, 119 and 69. [4]

4 A is a gas at room temperature; outline briefly one other method of measuring the relative molecular mass of A. [3]

5 Explain carefully how the pairs of peaks in the mass spectrum of compound B at 338 and 340, and at 269 and 271, arise. [2]

6 B is a covalent substance; suggest a structural formula which would be compatible with this mass spectrum. [2]

7 C is an isomer of B, and also covalent; suggest the mass of a peak which might occur in the mass spectrum of C, but not of B, explaining your reasoning. [2]

8 If you applied a method similar to that which you used in question 4, what value would you obtain for the relative molecular mass of compound B? [1]

9 Three methods of analysing hydrocarbons are: combustion, followed by measuring the masses of CO_2 and H_2O formed; the method you described in question 4; and mass spectrometry. Compare the information obtainable by these three methods. [3]

26 Analysis of Magnesium Hydroxide

The purity of a solid sample of magnesium hydroxide may be measured as follows. Weigh out a sample of impure magnesium hydroxide, and add it to exactly $50 \, cm^3$ of 1M hydrochloric acid. Warm the solution gently until all the solid dissolves. Then add a few drops of phenolphthalein indicator, and titrate with 1M sodium hydroxide solution until the solution turns a permanent pink colour.

In this titration the quantity of hydrochloric acid must be more than enough to react with all the magnesium hydroxide. The amount of unreacted hydrochloric acid is then determined by 'back titration' with the sodium hydroxide.

1 Write a balanced equation for the reaction of magnesium hydroxide with hydrochloric acid. [1]

2 Calculate the total number of moles of hydrochloric acid taken in this experiment. [1]

3 Remembering that some of this acid must be left over, suggest a sensible number of moles of magnesium hydroxide to be taken in this experiment. [2]

4 Hence suggest a suitable mass of magnesium hydroxide to be weighed out (A_r: H 1, O 16, Mg 24). [1]

5 What piece of apparatus would be suitable for measuring out the hydrochloric acid? [1]

6 What would you wash the burette out with before beginning the experiment? [1]

7 If the balance weighs the magnesium hydroxide to an accuracy of 0.01 g, would it be possible to determine the purity of the sample to the nearest 0.2%? Explain your answer. [3]

8 Suggest a practical reason why this technique of back titration might be better than simply titrating the magnesium hydroxide with standard hydrochloric acid. [2]

This method of analysis would work well if the magnesium hydroxide were contaminated with small amounts of an inert substance such as magnesium chloride, but not with some other contaminants.

9 What sort of contaminants would prevent this analysis from giving an accurate answer? [2]

10 Suggest a contaminant that could cause the purity to appear to be over 100%, and explain your answer. [2]

11 In each of the following cases, explain whether the experiment could be performed equally well with the alterations suggested:
a) using 0.5M sulphuric acid instead of 1M hydrochloric acid;
b) decreasing the concentrations of the solutions by a factor of ten, and using one tenth as much magnesium hydroxide. [4]

27 Analysis of Sodium Hydroxide

Samples of sodium hydroxide are often contaminated by sodium carbonate. The proportions of the two substances in a mixture can be measured as follows.

Take 0.1 g of the mixture and dissolve it in 25.0 cm^3 of water. 10.0 cm^3 of the solution is taken and titrated directly with 0.1M hydrochloric acid until the yellow colour changes to orange. A second 10.0 cm^3 of the solution is then taken in another conical flask, and 10 cm^3 of 0.05M barium chloride solution added, when a white precipitate forms. The solution is then titrated with 0.1M hydrochloric acid using phenolphthalein indicator. The solution loses its pink colour at the end point; the acid should be added slowly to prevent it reacting with the white precipitate.

The explanation of the experiment is as follows. In the first titration the acid reacts with both the sodium hydroxide and the sodium carbonate:

$$NaOH + HCl \rightarrow NaCl + H_2O$$

$$Na_2CO_3 + 2HCl \rightarrow 2NaCl + CO_2 + H_2O$$

In the second titration the sodium carbonate is removed from the solution by precipitation with excess barium chloride solution; the sodium hydroxide alone is then titrated with the acid. If the volumes of hydrochloric acid required for the two titrations are v and w cm^3, then the amounts of sodium hydroxide and sodium carbonate in the original sample are given by:

$$\text{Mass of NaOH} = 0.01 \times w \text{ g}$$

$$\text{Mass of Na}_2\text{CO}_3 = 0.013\,25 \times (v - w) \text{ g}$$

---

1 What reaction occurs between the sodium carbonate and the barium chloride?
[1]

2 Why should the acid be prevented from reacting with the white precipitate? [2]

3 What apparatus would be the simplest for measuring out (a) the 10 cm^3 of the sodium hydroxide/sodium carbonate solution and (b) the 10 cm^3 of barium chloride solution? [2]

4 What colour is methyl orange in acidic solution, and phenolphthalein in alkaline solution? [2]

5 Give ionic equations for the reactions of sodium carbonate and sodium hydroxide with hydrochloric acid. [2]

6 Is it necessary to weigh out the 0.1 g of the mixture accurately in order to find the proportions of sodium hydroxide and sodium carbonate? [1]

7 Prove that the masses of sodium hydroxide and sodium carbonate are as given by the formula at the end of the passage (A_r: H 1, C 12, O 16, Na 23). [4]

8 Suppose that your two titrations were 10.0 and 8.0 cm^3 respectively, each with an accuracy of ± 0.2 cm^3. Calculate the highest and lowest masses of sodium hydroxide and sodium carbonate that there could have been in the original sample. [4]

9 How might sodium hydroxide become contaminated with sodium carbonate? [2]

28 Analysis of Silver Alloys

Alloys of silver often contain the metals silver, copper and nickel. The alloys may be analysed in the following way.

Take about 1 g of the alloy, and add concentrated nitric acid. A reaction takes place, with nitrogen dioxide being evolved. When the metal has dissolved, evaporate the solution to remove most of the unreacted acid. Add water, and then sodium chloride solution. Boil the solution gently, away from strong light, and then filter, and wash the precipitate with distilled water. The precipitate of silver chloride is dried by heating in a crucible, and weighed.

The filtrate, which contains Cu^{2+} and Ni^{2+} ions, is neutralised with ammonium hydroxide until a slight precipitate is produced, and then treated with sulphur dioxide solution and excess ammonium thiocyanate, NH_4CNS. Copper is precipitated as CuCNS, filtered, washed, dried and weighed.

The filtrate is treated with a solution of dimethylglyoxime, $HO-N=C(CH_3)-C(CH_3)=N-OH$, and made just alkaline with ammonia solution. A precipitate of nickel(II) dimethylglyoximate, $(HO-N=C(CH_3)-C(CH_3)=N-O)_2Ni$, appears, and after warming for 20 minutes, is filtered, washed and dried.

---

1 Give the oxidation numbers of nitrogen in HNO_3 and NO_2. What is oxidised by the nitric acid? [3]

2 Give an equation for the reaction of silver nitrate with sodium chloride. Why should the precipitate be kept away from strong light? [2]

3 What can you conclude from this experiment about the solubilities of nickel(II) chloride and nickel(II) thiocyanate? Explain your reasoning. [2]

4 What is the charge on the thiocyanate ion? Hence give the name of CuCNS. What do you think the role of the sulphur dioxide solution might be in this reaction? [3]

5 What is the charge on the dimethylglyoximate ion? [1]

6 In an experiment, 0.962 g of an alloy was taken, and produced 0.817 g of CuCNS. What percentage of the alloy is Cu (A_r: Cu 63.5, C 12, N 14, S 32)? [3]

7 It is thought that the alloy might also contain some lead, as well as silver, copper and nickel. Given the solubilities of the compounds listed below, suggest how you might incorporate an analysis for lead into the scheme above, and explain your reasoning. [3]

	Ag	Cu	Ni	Pb
sulphide	insol.	insol.	insol.	insol.
sulphate	sol.	sol.	sol.	insol.
iodide	insol.	→$CuI\downarrow$ + I_2	sol.	insol.
carbonate	→$Ag_2O\downarrow$	insol.	insol.	insol.
chloride	insol.	sol.	sol.	insol.

8 Alloys can also be analysed by incorporating a very small sample of them into one of the electrodes of an arc, and then analysing the light emitted from the arc with a spectrometer; different metals give off light of characteristic frequencies. Suggest three advantages that this method would have in the industrial analysis of alloys over the precipitation methods. [3]

29 Preparation of Caesium Dichloroiodide

Compounds between the alkali metals and the halogens, with the metal and the non-metal in the ratio 1 : 1, are familiar and easy to form; sodium chloride, NaCl, is a simple example. There are, however, other compounds possible between the alkali metals and the halogens; one metal atom can be bonded to three non-metal atoms. These compounds contain the M^+ ion as usual, but the negatively charged ion contains three halogen atoms held together by covalent bonding. The preparation of such a compound is described below.

Caesium dichloroiodide, $CsICl_2$, contains the ion ICl_2^-, in which the iodine atom is bonded to two chlorine atoms; it is prepared as follows. 5 g of caesium chloride is dissolved in $50 \, cm^3$ of water, and solid iodine added. The mixture is heated to about 90 °C, and chlorine bubbled through the mixture, until the iodine just dissolves. The mixture is cooled, and the precipitated $CsICl_2$ is filtered off, washed with a little KI solution, and then water, and dried.

1 Suggest a precaution that should be taken when carrying out this experiment. [1]

2 What is the oxidation number of iodine in (a) I_2, (b) $CsICl_2$? What acts as an oxidising agent in this preparation? [3]

3 Give a balanced equation for the reaction. Hence calculate the mass of iodine that should be used in this experiment, if there is to be excess of neither caesium chloride nor iodine. What volume of chlorine (at room temperature and pressure) would be required? (A_r: Cs 133, Cl 35.5, I 127; 1 mole of any gas occupies $24 \, dm^3$ at room temperature and pressure.) [5]

4 What physical property of $CsICl_2$ differs from that of most Group I compounds? [1]

5 a) The precipitated $CsICl_2$ is washed with a little potassium iodide solution; what substance might be dissolved by this, but not by water? [1]
   b) Suggest one other substance which might then be removed by washing with water. [1]

6 How many electrons are there in the outermost shell of the iodine atom in (a) an I^- ion, (b) an ICl_2^- ion? [2]

7 Do you think that fluorine would form a corresponding compound $CsFCl_2$? Explain your reasoning. [2]

8 When caesium dichloroiodide is heated, a gas is given off, and a white solid remains. Three different ideas are proposed as to what is happening: the gas could be chlorine, and the solid caesium iodide; the gas could be iodine monochloride (ICl) and the solid caesium chloride; or a mixture of the two gases could be formed with a mixture of the two solids. Suggest how you could investigate in a simple experiment which of these suggestions is correct. [3]

9 In fact the gas is ICl, and the solid formed CsCl; suggest why this product might be formed in preference to the other. [1]

30　Chrome Alum

The alums are a series of double salts of formula $M^+M^{3+}(SO_4)_2 \cdot 12H_2O$; they contain one anion, but two different cations. Chrome alum has the formula $KCr(SO_4)_2 \cdot 12H_2O$; in aqueous solution it behaves like a mixture of potassium sulphate and chromium(III) sulphate, but when crystallised it has a definite composition, with a K : Cr ratio of 1 : 1. Many different alums are known; M^+ can be ammonium, or any of the alkali metals except lithium, and M^{3+} can be iron(III), chromium(III) or aluminium. The alums are isomorphous, that is they have the same crystal structure, and one alum can be made to grow on the crystal of another.

Chrome alum is prepared as follows. $5\,g$ of $K_2Cr_2O_7$ is dissolved in $16\,cm^3$ of water, and $5\,cm^3$ of conc. sulphuric acid added carefully with stirring. The potassium dichromate dissolves, and the solution is cooled to $30\,°C$. $3\,cm^3$ of ethanol is added in drops; the temperature is kept below $50\,°C$. The solution turns green, and is left in an evaporating basin to crystallise; $9\,g$ of violet crystals are formed.

The alum can be analysed for chromium as follows. $2\,g$ of chrome alum is added to $100\,cm^3$ of water, and warmed to dissolve it. Dilute ammonium hydroxide solution is added, forming a green-grey precipitate; addition continues until no further solid is formed. The solution is then boiled to remove excess ammonia, in which the chromium(III) hydroxide is slightly soluble. The precipitate is filtered, washed with water, and transferred to a weighed crucible, where it is heated strongly to constant mass; this converts the chromium(III) hydroxide to chromium(III) oxide.

1　What is the formula of ammonium alum, ammonium aluminium sulphate?　[1]

2　Would you expect a mixture of magnesium and copper(II) sulphates to form an alum $MgCu(SO_4)_2 \cdot 12H_2O$ isomorphous with chrome alum? Explain your reasoning.　[1]

3　In what way does chrome alum differ from a powdered mixture of chromium(III) sulphate and potassium sulphate?　[2]

4　Give the colours of chromium(III) sulphate and of chrome alum.　[2]

5　Suggest a reason why lithium alums cannot be formed.　[1]

6　What is the oxidation number of Cr in $K_2Cr_2O_7$? What is the role of the ethanol in this preparation, and to what might it be converted?　[3]

7 Calculate the number of moles of $K_2Cr_2O_7$ taken in this experiment, and hence calculate the percentage yield obtained. Assume that the sulphuric acid and ethanol are in excess (A_r: K 39.1, Cr 52, O 16, S 32, H 1).
Suggest one reason why the yield is less than 100%. [4]

8 How is the ratio of K:Cr fixed as 1:1 in this preparation? [1]

9 When the chrome alum is analysed, ammonium hydroxide is used rather than sodium hydroxide, as sodium hydroxide would cause the precipitate to dissolve. What property is chromium(III) hydroxide showing here? [1]

10 What is the green-grey precipitate (para. 3)? [1]

11 Suggest two substances which are removed when the chromium(III) hydroxide is washed with water. [2]

12 Give an equation for the thermal decomposition of chromium(III) hydroxide. [1]

31 Silver(II) Compounds

Silver is a member of the second transition series; its atoms have eleven electrons outside the inert gas configuration of krypton, and these are arranged $4d^{10}5s^1$. The most common oxidation state of silver is $+1$, but some compounds in higher oxidation states are known. The Ag^{2+} ion is not stable on its own in aqueous solution, but can be stabilised by complex ion formation.

An example of the preparation of a silver(II) compound is as follows. A solution of silver(I) nitrate in ice-cold water is prepared and pyridine, C_5H_5N, is added. A solution of potassium peroxodisulphate, $K_2S_2O_8$, is then stirred in slowly, while the mixture is kept cool. The solution is then filtered, and the precipitate washed with a little water, and dried. Orange crystals of formula $[Ag(C_5H_5N)_4]S_2O_8$ are formed; they decompose slowly over several days.

A more stable complex can be prepared by stirring the orange crystals into ice-cold water, and adding a solution of pyridine-2-carboxylic acid, a bidentate ligand:

After 30 minutes a brown precipitate can be filtered, washed and dried; its composition is $Ag(C_5H_4N-COO)_2$.

---

1 Give the electron configuration of the Ag^+ and Ag^{2+} ions. [2]

2 Suggest why a solution of silver(I) nitrate is colourless, but the two silver(II) complexes prepared above are orange and brown respectively. [2]

3 Suggest which element would be most likely to form a silver(II) compound when added to AgCl. [1]

4 a) The peroxodisulphate ion is reduced to SO_4^{2-} in this experiment; how many electrons are required to reduce one peroxodisulphate ion? [1]
　 b) Why then do 2 moles of Ag^+ require 3 moles of peroxodisulphate for complete reaction? [2]

5 The pyridine molecule contains a six-membered ring. Which atom do you suppose is bonded to the silver(II) ion? Explain your reasoning. [2]

6 The coordination number of the silver ion is the same in the two complexes; what is its value? [1]

7 Explain the meaning of 'a bidentate ligand' (para. 3). [1]

8 a) Sketch the structure of the complex $Ag(C_5H_4N-COO)_2$ [2]
   b) What is the charge on C_5H_4N-COO? [1]

9 Suggest what ions would be present in the filtrate when the brown solid is filtered off (para. 3). [2]

10 Suggest a suitable mass of pyridine-2-carboxylic acid to be added to 1 g of $[Ag(C_5H_5N)_4]S_2O_8$ in the second preparation (para. 3). Express your answer to 1 significant figure only. (A_r: Ag 108, S 32, O 16, N 14, C 12, H 1) [3]

32 Preparation of Pentan-1-ol

Esters may be thought of as consisting of two parts, one derived from a carboxylic acid, and one from an alcohol; thus ethyl pentanoate is the ester of pentanoic acid and ethanol. If this ester is reduced, then it splits up giving ethanol, and pentan-1-ol which would be obtained if pentanoic acid were reduced. This reduction is conveniently performed by sodium and ethanol.

10 g of ethyl pentanoate is placed in a flask fitted with a reflux condenser, with $150\,cm^3$ of ethanol, which must have been dried carefully. 19 g of sodium is added in small pieces, over a period of about 20 minutes. The mixture is then heated for an hour, when all the sodium reacts. The flask is then fitted with a fractionating column, and about $50\,cm^3$ of ethanol is boiled off. $60\,cm^3$ of water is then added, and the distillation continues until the temperature at the top of the column reaches 83 °C, when virtually all the ethanol has been removed. The mixture is then steam distilled, with about $40\,cm^3$ of distillate being collected. The pentan-1-ol, a colourless liquid, is separated, dried over anhydrous calcium sulphate, and distilled, with the fraction boiling at 137 °C being collected. The yield is about 5 g.

---

1 Give the structural formula of ethyl pentanoate. [1]

2 a) Give the formulae of two compounds, apart from esters, which could be reduced to form pentan-1-ol. [2]
  b) Could pentan-2-ol be formed by reduction of an ester? Explain your answer. [2]

3 Give a balanced equation for the reduction of ethyl pentanoate with sodium and ethanol, forming pentan-1-ol and sodium ethoxide (C_2H_5ONa). [2]

4 a) Calculate the number of moles of sodium and of ethyl pentanoate taken, and hence show that the sodium is in excess; what happens to the excess sodium? [4]
  b) What is the final percentage yield (A_r: Na 23, H 1, C 12, O 16)? [2]

5 a) Why is a fractionating column used in the removal of ethanol, rather than just a condenser? [1]
  b) Is the ethanol (b.pt. 78 °C) taken off the top or the bottom of the fractionating column? [1]

6 Why does pentan-1-ol have a higher boiling point than ethanol? [1]

EXPERIMENTAL SITUATIONS

7 The mixture of pentan-1-ol and water is steam distilled; what does this tell you about the miscibility of pentan-1-ol and water? How are the pentan-1-ol and water separated? [2]

8 What happens to the anhydrous calcium sulphate when it acts as a drying agent? How can it be separated from the pentan-1-ol? [2]

33 Friedel-Crafts Reaction

The Friedel–Crafts reaction can be used to introduce side chains into a benzene ring, including the formation of a new $C-C$ bond. The reaction involves a benzene compound and an acyl chloride, in the presence of aluminium chloride as a catalyst. The role of the aluminium chloride is not just that of a catalyst; it also forms a complex with the reaction product, and so slightly more moles of aluminium chloride are required than moles of product are expected.

The ketone $C_6H_5COCH_3$ is synthesised by mixing 30 g of benzene and 20 g of anhydrous aluminium chloride in a flask fitted with a reflux condenser. The flask stands in a bath of cold water while 10 g of ethanoyl chloride, CH_3COCl, is slowly added over a period of about 30 minutes. When all the ethanoyl chloride has been added the mixture is kept at 50 °C for 1 hour, and then poured into 200 cm³ of cold water. Heat is evolved, and an oil separates on the surface of the water. A little solid remains undissolved, and this is dissolved by the addition of conc. HCl.

The mixture is then placed in a separating funnel, and the aqueous layer discarded; the organic layer is washed with water, sodium carbonate solution, and then water, and left to stand over anhydrous calcium chloride. The benzene is then removed by distillation, and the product is distilled off at about 201 °C.

1 Give a balanced equation for the Friedel-Crafts reaction described above.
[1]

2 Suggest and explain two precautions which would need to be taken in this reaction.
[2]

3 Suggest why heat might be evolved when the reaction mixture is added to cold water (para. 2).
[2]

4 What might the solid be which remains undissolved (para. 2), and why should it dissolve in hydrochloric acid?
[2]

5 Why is the crude product washed with sodium carbonate solution? Why is it made to stand over anhydrous calcium chloride (para. 3)?
[2]

6 For ketones of higher relative molecular mass, the boiling point is higher than that of $C_6H_5COCH_3$; decomposition can be a problem in the final distillation. How can this problem be overcome?
[1]

7 Calculate the numbers of moles of aluminium chloride and ethanoyl chloride taken in this experiment, and explain their relative magnitudes (M_r: ethanoyl chloride 78.5, aluminium chloride 133.5). [2]

8 Calculate the number of moles of benzene taken, and explain what role it has, other than as a reactant (M_r: benzene 78). [2]

9 What starting materials would you take if you wanted to prepare (a) $C_6H_5-CO-C_2H_5$ and (b) $C_6H_5-CO-C_6H_5$ by the Friedel-Crafts reaction? Could pentan-3-one be made in this way? [5]

10 Suggest a problem that would arise if this method were extended to the reaction of ethanoyl chloride with methylbenzene. [1]

The enthalpy changes of many reactions cannot be easily measured directly because the reactions do not occur at room temperature; others are difficult because the reactions occur much too slowly. In these cases the enthalpy change for a reaction may often be measured indirectly, by measuring the enthalpy changes for other reactions in which the reactants and products take part.

An example of this is the reaction of CaO and CO_2 giving $CaCO_3$. Calcium oxide does absorb carbon dioxide, but only very slowly, while calcium carbonate decomposes, but only on very strong heating. But calcium oxide and calcium carbonate will both react readily with dilute hydrochloric acid, and the enthalpy changes for these two reactions can be easily measured. A Hess cycle then gives the enthalpy change for the reaction of calcium oxide and carbon dioxide:

$$CaO\,(s) + CO_2\,(g) \longrightarrow CaCO_3\,(s)$$

$+2HCl$ \hspace{3cm} $+2HCl$

$$CaCl_2\,(aq) + H_2O\,(l) + CO_2\,(g)$$

The reaction can be carried out in a small plastic beaker. $50\,cm^3$ of 0.4M hydrochloric acid are added to the beaker, and the temperature measured. A suitable amount of calcium carbonate is then added, and the mixture stirred. When the reaction is over, the temperature is again noted. Knowing the specific heat capacity of the solution, the heat evolved is calculated. The experiment is then repeated using calcium oxide in place of calcium carbonate.

1 Why can the enthalpy change for the reaction of CaO and CO_2 forming $CaCO_3$ not be obtained by measuring directly the heats of formation of CaO, CO_2 and $CaCO_3$? [1]

2 Why is a plastic cup used in this experiment rather than a glass beaker? [1]

3 Suggest an important source of error in this experiment. [1]

4 A further error can arise from the fact that CaO can absorb water slowly from the atmosphere. Suggest what it would form, and what precaution could be taken to overcome this problem if only an old sample of CaO were available. [2]

5 What is the maximum amount of calcium carbonate that could react with the amount of acid taken (A_r: Ca 40, O 16, C 12)? [2]

6 If this amount of calcium carbonate is taken, the temperature rise is about $1\,°C$, which is too small to be measured very accurately. In each of the following cases, explain whether the proposed change in conditions would result in a larger temperature change.
a) Use $100\,cm^3$ of 0.4M HCl, and the same amount of $CaCO_3$.
b) Use $50\,cm^3$ of 0.4M HCl, and twice as much $CaCO_3$.
c) Use $50\,cm^3$ of 0.8M HCl, and the same amount of $CaCO_3$. [3]

7 a) Using the data above, and the fact that it takes 4.2 J to heat $1\,cm^3$ of aqueous solution through $1\,°C$, calculate the amount of heat evolved in this experiment. (Assume the total volume of the solution at the end is $50\,cm^3$.)
Hence calculate the amount of heat evolved per mole of $CaCO_3$ taken. [2]
b) If the temperature rise recorded is really $1 \pm 0.5\,°C$, give the error limits to your value for the heat evolved per mole. [1]

8 When the experiment is repeated using 0.56 g of CaO, the temperature rise is $8\,°C$. Hence calculate the amount of heat evolved per mole of CaO. [2]

9 Use your answers to questions 7 and 8, and the cycle above, to calculate ΔH for the reaction $CaO + CO_2 \rightarrow CaCO_3$.
Assuming the error in temperature measurement in each experiment to be $\pm0.5\,°C$, give the possible errors in your value of ΔH. [3]

10 The value of ΔH from a Data Book is actually $-178\,kJ\,mol^{-1}$; explain why your answer is numerically too small. [1]

11 Magnesium carbonate decomposes at a lower temperature than calcium carbonate; what does this suggest about the value of ΔH for the decomposition of magnesium carbonate? [1]

35 The Dimerisation of Ethanoic Acid

When ethanoic acid dissolves in some organic solvents, its molecules pair up by forming hydrogen bonds to each other, giving a dimer:

$$2CH_3COOH \rightarrow CH_3-C \begin{matrix} {}^{O---H-O} \\ {}_{O-H---O} \end{matrix} C-CH_3$$

It is possible to show that this is happening by investigating the way in which ethanoic acid distributes itself between two immiscible solvents.

If a solute is added to a mixture of two immiscible solvents A and B, then two layers are formed, one a solution of the solute in A, and the other a solution of the solute in B. The concentrations of the solute in the two layers are normally related:

$$K = [solute]_A / [solute]_B \qquad (1)$$

where K is an equilibrium constant called the partition coefficient.

However if the solute forms dimers in solvent A, then it is possible to show that

$$K = [solute]_A / [solute]_B^2 \qquad (2)$$

where K is a (different) constant. By dissolving ethanoic acid in water and tetrachloromethane, and then varying the amounts of the acid taken, it is possible to find whether equation (1) or (2) holds in this case, and hence whether the ethanoic acid is dimerised when dissolved in tetrachloromethane.

$25 \, cm^3$ of water is added to $25 \, cm^3$ of tetrachloromethane, and a suitable mass of ethanoic acid is added. The mixture is shaken, and then allowed to separate into two layers. $10 \, cm^3$ of the water layer is removed by pipette, phenolphthalein indicator added, and the solution titrated with 0.1M NaOH solution. $10 \, cm^3$ of the tetrachloromethane layer is also removed by pipette, phenolphthalein and $20 \, cm^3$ of water are added, and the solution is titrated with 0.02M NaOH. The solution needs constant shaking during the titration to ensure that all the acid is extracted into the aqueous layer. The values of K given by equations (1) and (2) are calculated. The experiment is then repeated with a different amount of ethanoic acid; the new values of K are calculated, and compared with the previous values.

1 a) Is it necessary to know the exact amount of ethanoic acid taken in the experiment?
   b) Is it necessary to take exactly $25\,cm^3$ of water in the experiment?
   c) Is it necessary to take exactly $10\,cm^3$ of the aqueous layer in the experiment?
   In each case explain your reasoning. [3]

2 In an experiment a solution of 0.2M NaOH solution is taken in error instead of a 0.1M solution. Will this affect the final conclusion of the experiment? Explain your reasoning. [2]

3 This question is designed to estimate a suitable quantity of ethanoic acid to take in this experiment.
   a) Suggest a suitable volume for the titration of the aqueous layer.
   b) Calculate the number of moles of acid therefore needed in the $10\,cm^3$ sample, and hence the number of moles in the whole aqueous layer.
   c) Assuming that most of the acid is present in the aqueous layer, estimate a suitable volume of ethanoic acid to take (for ethanoic acid $M_r=60$, and density $= 1.05\,g\,cm^{-3}$). [4]

4 Suggest a reason why 0.02M NaOH is used in the titration of the organic layer, rather than 0.1M NaOH. [1]

5 Towards the end of the titration of the tetrachloromethane layer, when NaOH is added the aqueous layer turns pink, while the tetrachloromethane layer remains colourless. After shaking both layers are colourless. Explain why this is observed. [3]

6 In an experiment it is found that the volume of NaOH needed to neutralise the tetrachloromethane layer is $1.7\,cm^3$, which is rather too small to be measured accurately. Suggest two simple steps which could be taken to increase the volume measured (other than using more ethanoic acid). [2]

7 Does the fraction of ethanoic acid in the aqueous phase increase, decrease or stay the same when more ethanoic acid is used (a) if equation (1) is true and (b) if equation (2) is true? [2]

8 This method cannot be used to investigate whether ethanoic acid is dimerised in ethanol; suggest a reason why not. [1]

9 Suggest a reason why ethanoic acid should be dimerised in some organic solvents, but not in aqueous solution. [2]